THE LAST BUILDING AT
PACKERTON YARDS

Louis A. Robertella

ISBN:10:1530629462
ISBN-13:978-1530629466

DEDICATION
To those who desire to pass on a memory of the past to the future.

CONTENTS

ACKNOWLEDGMENTS

In memory of Asa Packer, and the many men and women who helped build the Lehigh Valley Railroad and the Bethlehem Steel Company, two industries which changed the history of America.

1-INTRODUCTION

On an excursion in 2003 to take photographs of the historic town of Jim Thorpe which was formerly known as Mauch Chunk, Pennsylvania, I happened to notice a large deserted building off in the distance. This building was in an open field on my right as I approached the town. I made a detour into the field to have a closer look at the building.

Crossing several sets of railroad tracks I approached the building which was in a deteriorated condition. I entered the structure and soon noticed that the construction of this building used a building technique that I did not expect in a building of this age.

Having been in many mill buildings of various types of approximately this age and earlier, I expected to see large wooden timbers and cross beams. This building had iron or steel columns and cross beams.

I took a number of photographs and decided to return and take measurements of the structure as I felt the building would not last much longer in the condition it was in.

I returned to the building several times during 2003 to complete the measurements. I had decided that this building would be an interesting structure to re-create using a computer design program that I had used for other studies.

Because of several projects I was then working on, I put aside this project for another time.

Eleven years later, in 2014 and with more time available, I looked through some uncompleted projects and found the measurements and photographs I had taken in 2003 of this unusual building. I decided to complete the project and finish the research I had started on this building.

I soon learned the building, after many attempts to restore it for various purposes, was demolished in 2007. Its fate had been decided through the courts. The building no longer existed.

I felt this structure would be interesting to virtually recreate as a 3D computer model.

The following study of this building was begun in 2015, eight years after its demolition.

What I learned of its history and use is presented in the following pages.

EARLY STEEL BUILDINGS

At the end of the nineteenth century, steel began to be used in the construction of large commercial buildings. This became the new method which allowed structures to be built higher and stronger than ever before.

Steel beams of various shape and sizes were being manufactured by steel producers as the demand for these structures increased. Steel rather than iron or cast iron which was brittle in comparison, made this transformation possible.

Steel beam structures at this time were assembled by using rivets to hold the various pieces of the infrastructure together.

Red-hot rivets were placed in pre-drilled holes in the steel beams. One man held the hot rivet in place with a tool designed for that purpose, while another man used a hand-held pneumatic hammer to hammer down the protruding end of the rivet into a rounded head, holding the pieces together.

This technique was used in the formation of steel structures that led to the construction of larger and taller buildings that we know as "Sky-Scrapers" today.

The first building using this new steel technique was the Home Insurance Company building in Chicago, built in 1885. This building used a steel and cast iron infrastructure, with the outside masonry providing the stability and not the steel core itself. The Home Insurance building was ten stories rising to 138 feet in height. The use of steel in structures also reduced the risk of a spreading fire. The Home Insurance Building was demolished in 1931.

HOME INSURANCE BUILDING. CHICAGO-1885.

THE RAND McNally BUILDING-CHICAGO 1890.

The first all steel framed building was the Rand-McNally Building built in Chicago in 1890. This building used the steel frame itself for stability and masonry as external façade. The Rand-McNally Building was demolished in 1911.

THE TOWER BUILDING. NEW YORK CITY 1889.

The first true "Skyscraper" built in New York City was the Tower Building on lower Broadway, built in 1889. It is considered the first skyscraper because it had a steel skeleton at its core and this steel skeleton supported the entire structure. The Tower building was eleven stories high. It was demolished in 1913.

Previous to the "Tower" building, structures in New York City were mostly five to seven stories in height and built of wood with brick exteriors. The use of steel as the inner skeleton is what gave new construction the ability to grow upwards creating that magnificent visage we know today as the New York City skyline.

THE FLATIRON BUILDING. NEW YORK CITY 1902
(Courtesy Library of Congress)

The famous Fuller or "Flatiron" Building built in New York City in 1902 was designed in 1898. This building used steel as the inner framework and main support and exterior masonry as ornament reflecting the style of the time.

Beside taller buildings being designed, all the new bridges built in Manhattan and elsewhere after this point, including the George Washington Bridge started in 1927, were constructed using riveted steel structures.

These riveted steel structures include such famous buildings as the Chrysler Building (1930), Empire State Building (1931), and most skyscrapers constructed by the mid 1900's.

THE PACKERTON YARD BUILDING

The last remaining building of the Lehigh Valley Railroad's Packerton Yards,
Packerton, Pennsylvania.

Asa Packer, a self-made man, opened the Lehigh Valley Railroad on Sept. 12, 1855 with the idea of transporting coal mined in the mountains of Eastern Pennsylvania to the many markets on the Atlantic seaboard.

His plan was to move anthracite coal from the mountains of Carbon County Pennsylvania to the coal chutes in the city of Easton, Pennsylvania. At Easton the coal was loaded into coal chutes which were used to fill coal barges using the Pennsylvania Canal which ran from Easton to Philadelphia, Pennsylvania.

Directly across the Delaware River from Easton is Phillipsburg, New Jersey. Phillipsburg was a hub for both the New Jersey Railroad, and the Morris Canal, a canal system in New Jersey.

At Phillipsburg the coal cars were transferred to the New Jersey Railroad system and coal was also loaded into barges on the Morris Canal which brought the coal to piers in Jersey City, New Jersey located opposite New York City on the Hudson River.

Coal was the main form of heating for residential and commercial buildings. Coal was used in locomotives which was the new mode of transportation at that time. There was also a huge demand for coal by the steamships of various kinds which all used steam engines for power.

Railroads provided fast, efficient transportation of both passengers and freight and made thousands of railroad jobs available to the growing populations in the east.

At the town of Burlington, Mahoning Township, Pennsylvania, Asa Packer purchased 47 acres of land in 1863 for a rail yard to expand operations for his Lehigh Valley Railroad. The town of Burlington was renamed Packerton and the rail yard was named Packerton Yards.

This area is located between the towns of Jim Thorpe (formally Mauch Chunk) and Lehighton with the Lehigh River as its eastern border. Packer wanted this location to serve as his headquarters for "all coal passing east

Packer's Lehigh Valley Railroad built a large rail and switching yard which had turntables to turn locomotives to specific tracks, coal car building shops, engine and car repair shops, an engine house for twenty-nine engines, warehouses, and switching yards.

Asa Packer became very wealthy from his railroad and other enterprises. He was also one of the founders of the Bethlehem Iron and Rolling Mills which later changed its name to Bethlehem Steel in 1899.

In 1865 he donated land and money for the Lehigh University at Bethlehem Pennsylvania, presently one of the best engineering universities in America.

Packerton Yards became a very busy site. In the year 1890 for example, a total of 792,765 loaded coal cars were received at the yard, an average of 3000 per day not including Sundays and holidays. (Railway World-1891)

The Lehigh Valley Railroad continued expanding for years until the slow demise of the Railroad Industry began in the 1950's.

Sometime after 1953 Packerton Yards, which handled mostly coal shipping was no longer used. All the buildings became dilapidated and were eventually torn down, except for one.

The last surviving building of the Packerton Yards, the subject of this study, is sometimes listed as a "Coal car repair" shop. In other mentions it is listed as a warehouse.

I found no indication, from markings on the floors of this building, or extra supporting features, etc. that this building once housed machinery large enough to repair coal cars or engines. There are no standard railroad tracks leading into the building since it is elevated sixty-four inches above the common ground level.

I concluded this building must have been used as a parts warehouse. There was a narrow gauge track, running down the center of the second floor which indicated that rolling platforms were used to move and retrieve items.

On the first floor some narrow track exists in various locations but may have been put in later since they are not as well constructed as the track is on the second floor.

Years after the building was no longer of use, a company that manufactured mobile homes used part of the rear of the building for that purpose. They made some renovations to the north end wall of the building but they too eventually closed operations.

In a "Morning Call" (an Allentown, Pennsylvania newspaper), article from 2006, when demolition of the building was being proposed, the article stated:

"Structure was used as a storehouse with two or three employees. The information is validated by a Lehigh Valley Railroad track map from 1917. But the overall Packerton site was used as a transfer station for freight cars from the 1870s until 1971."

Finding this information confirmed my original analysis regarding this building's use as a warehouse.

This last building at Packerton Yards was abandoned from the late 1950's till the yard was formally closed March 16, 1973.

PHOTO OF THE LAST BUILDING STANDING AT PACKERTON YARDS.
(Library of Congress photo-image dated 1979.)

Construction on this building was started in 1899 and was completed by 1905.

The buildings two floors combined equaled 42,000 square feet of usable space.

This last building was finally demolished in 2007 surviving for approximately 102 years.

At the time when I first noticed this building and took photographs of it, twenty-three years after the above photograph was taken, most of the windows were missing or broken beyond repair, and most of the steel structure was rusted.

The plaster coating used over the ceiling fire bricks and walls had fallen from the structure in many places.

The following illustrations made from computer models created for this project show how the building was constructed using the new steel technology of that time. It shows the interior walls as brick before they were plastered.

The illustrations used in this study make use of various colors to make particular parts more identifiable and to show with better clarity the various parts used in its construction.

2-EXTERIOR

The exterior of this building was constructed of brick walls between vertical steel columns which were connected by the framework to the entire superstructure of the building.

In the west wall of the building were twenty-one large windows on the second floor and sixteen double windows on the first floor. Also on the west wall were three large double doors that opened to the first floor. There were also two normal sized doors with a small window next to each door that open into stairwells, being Stairwells-1 and 2.

The east and west walls had identical windows and openings for large double doors. The east wall did not have any small doors or windows since there were no stairwells on the east side of the building. There was a combined total of forty-two windows on the second floor and thirty-two windows of the first floor in the east and west walls.

The north and south end walls were similar in that both had large double doors that opened to the center of the first floor. In the rear or north wall there were two normal doorways, one opened into a small office area, Stairwell-3 and the first floor. The other doorway opened into the first floor only.

The north end had a different window arrangement than the south end of the building. At some later period, the rear wall was modified; the views in this study show the north and south walls as originally constructed.

At a later date a large air compressor was put into this building. Piping in the building does not indicate that it had air pressure pipes serving it. They may have been removed but that is not indicated. The compressor must have been for air pressure used at different locations within the railroad yard. Both items were removed in later years.

What remains on the exterior of the building is a smokestack that goes into the ground next to the first floor and into the area where the boiler was located. It would seem that a boiler for steam pressure to power the compressor was the reason for the smokestack, no other reason for its use can be found.

The smokestack was constructed high enough and with a spark suppression device to indicate that it released heated smoke and ash into the air. A wooden stand was constructed around the top of the smokestack for maintenance purposes.

There were remains of a few downspouts for the removal of rain water from the roofs but they were beyond physical reach for measurement when this project was started. As shown in the following examples, the exterior of the building appears as when it was first built.

This is a view of the entire building showing the western side as it would have appeared in 1905. The Lehigh River was located on the eastern side or behind the building as shown in this view. This view also includes the smokestack which was added at a later time and existed until the building was demolished in 2007. Photo 1.

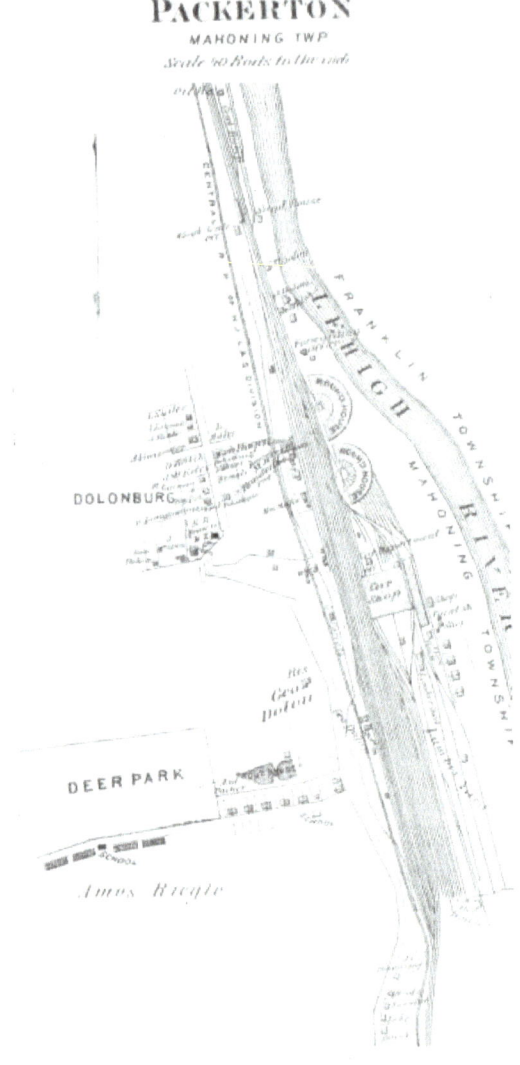

This 1875 Beer's Map of Packerton shows Packerton Yards and the Lehigh River flowing south past the rail yard towards Easton, Pennsylvania.

This view shows the south end of the building as originally constructed. A ramp led up to the first floor level of the building on the west side. To the left of the large center doors on the south wall (foreground), is a mail or order slot that used a spring return to close the door. Running down the center top of the building is the clerestory, used to admit daylight into the building on the second floor.

This view of the south end of the building shows the smokestack in place. The smokestack was added at a later date. This smokestack was in place when the building was demolished in 2007. Photos 2, 2A, 2B, 2C, 2D.

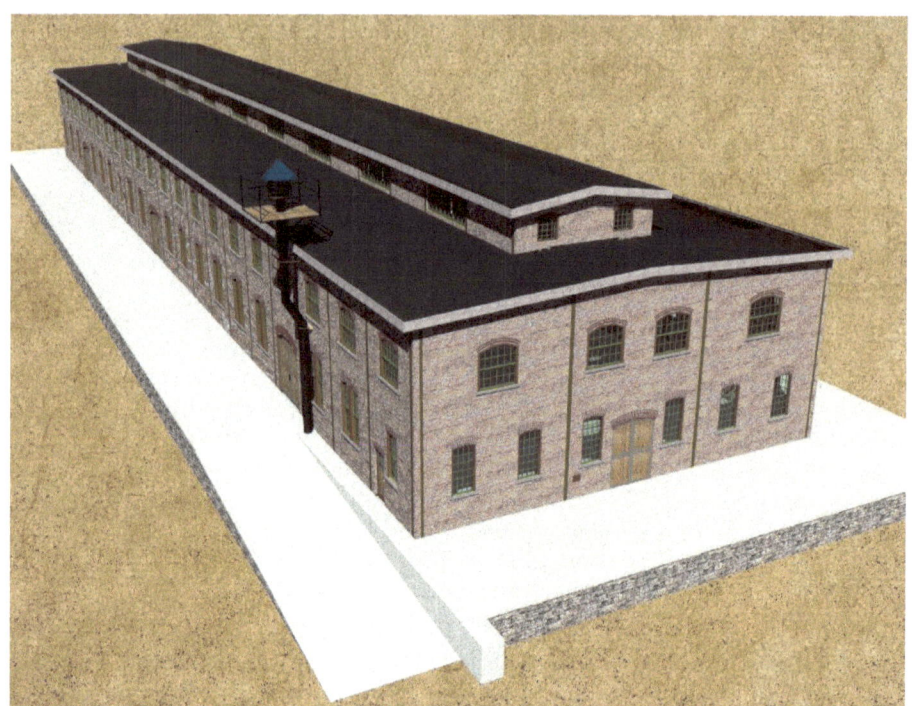

This view from a higher angle shows the asphalt roofing and the wooden base built for maintenance of the smokestack.

This view shows the north end of the building. Note the different type and window arrangement when compared to the south end. The small door on the right opened to a small office, the first floor, and Stairwell-3. The small door on the left opened to the first floor. Photos 3, 3A, 3B, 3C, 3D, 3E.

This view, looking southeasterly, shows the smokestack near the southerly end of the building. The small door in the distance is the entrance to Stairwell-1. Photos 4, 4A, 4B.

Top view of the entire building and its platform which is raised 64" above the common ground level.

The outer walls of the building, less the roof overhang and clerestory, measure 320' X 72'.

Approximate height of the building was 42 feet.

Total square footage of usable space was 42,000 Sq. Ft.

Along the east side of the building (top of image) a railroad track allowed railroad cars to deliver or receive materials from the building.

3-CONSTRUCTION

This building was built using a steel frame with brick outer walls. The outer walls were divided into sections between vertical steel beams. The ceilings used steel arches designed to hold fire bricks set on steel rails as a way to prevent a fire on the first floor from spreading to the second.

On the second floor there was a full clerestory monitor and additional walls were constructed having fire doors to divide the space on the second floor into sections. In the event of fire, these walls would help prevent the fire from spreading to other sections.

Each of three stairwells were constructed as separate bricked areas within the main building and would also keep fire from spreading to the second floor or roof.

Overall dimensions for the building were 320 feet in length, 72 feet in width, and approximately 42 feet in height.

The building was constructed on a raised base being 64 inches above ground level. The base was approximately 360 feet X 112 feet X 64 inches.

The base was constructed first then a cement pad area was poured for the actual placement of the steel superstructure.

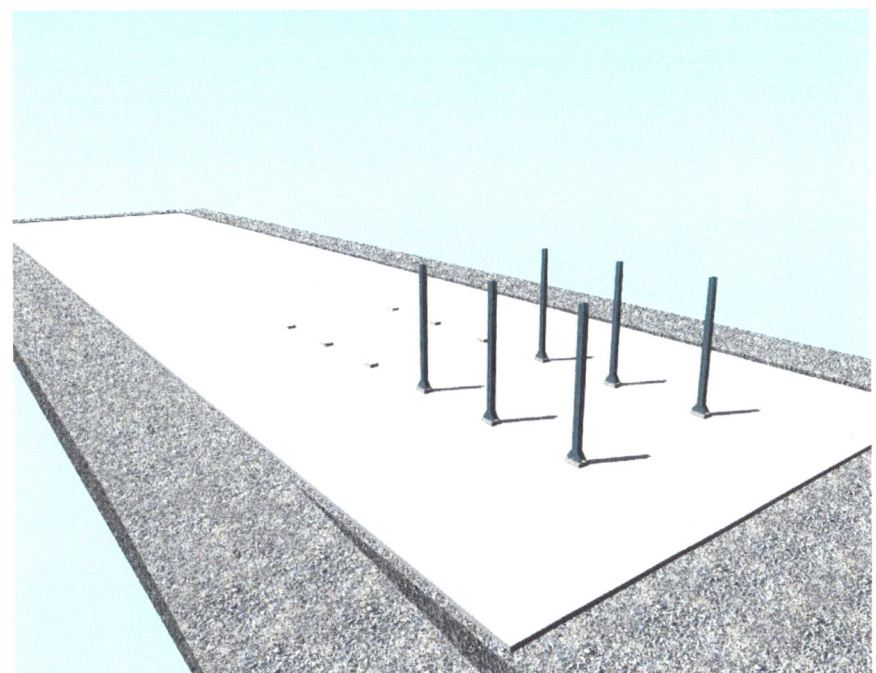

Riveted steel beams measuring 9-3/8 inches X 10-1/4 inches X 16 feet were positioned down the center of the structure. The steel used was 3/8" thick. The distance between each row was twenty-three feet. Each column in each row was spaced fifteen feet apart.

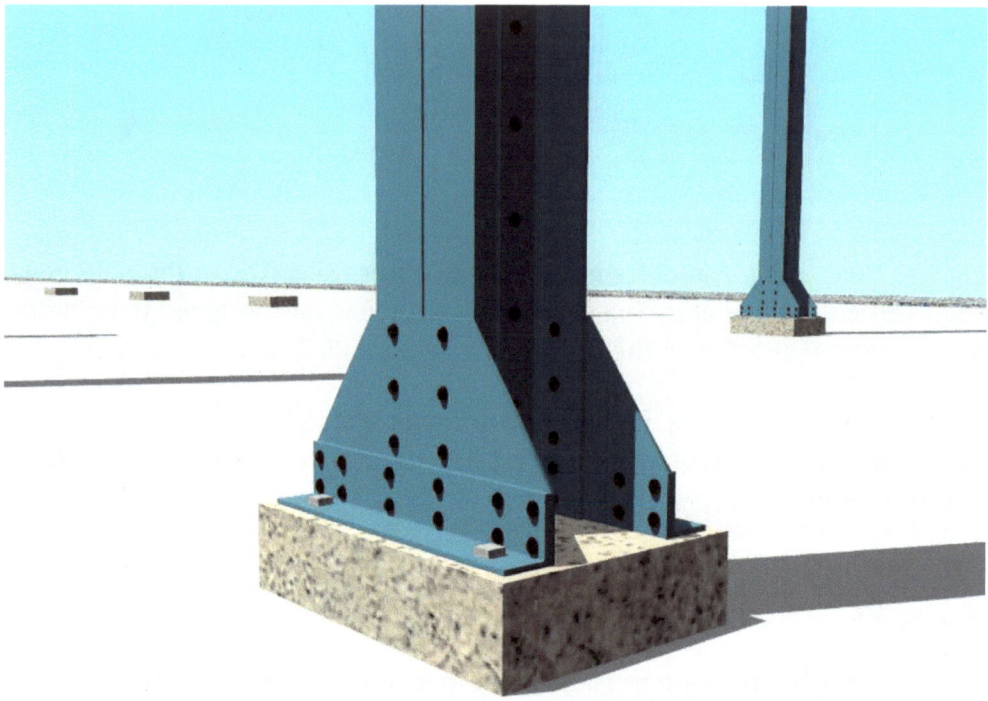

Each column was bolted to a column base. Note the rivet assembly. Photo 5

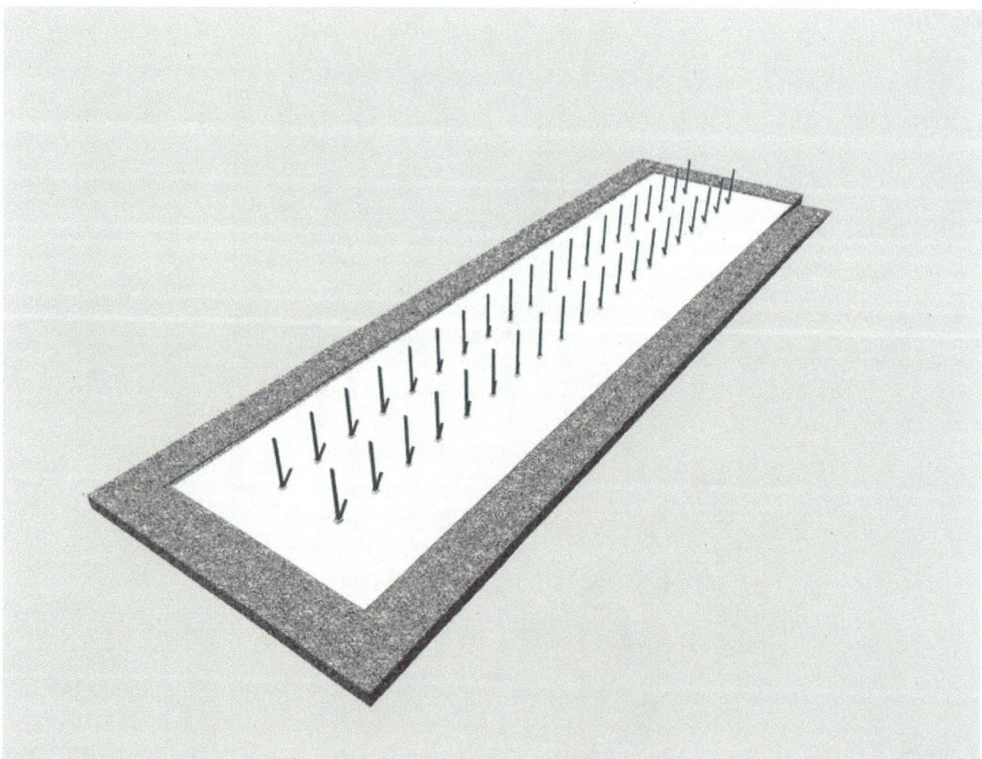

This image shows the forty central columns which formed the central section of the superstructure. Columns in each row are fifteen feet apart. The distance between each row of columns is twenty-three feet.

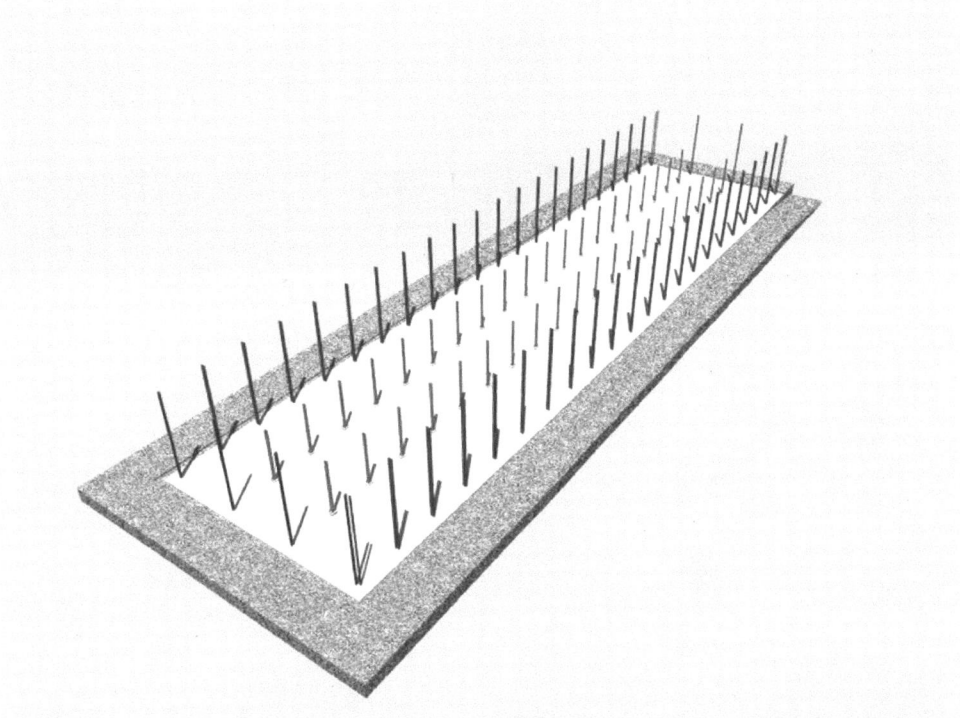

Around the perimeter of the central columns (blue) were placed steel columns (brown) between which the brick outer walls were built.

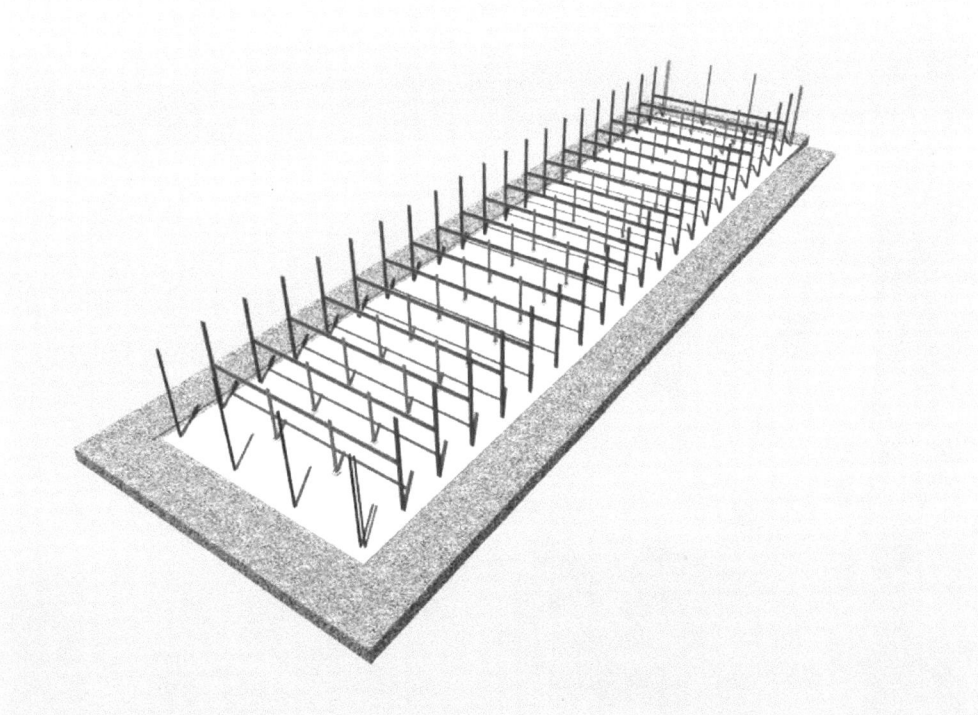

Beams spanning between central and outer columns are 18" X 6" steel beams.

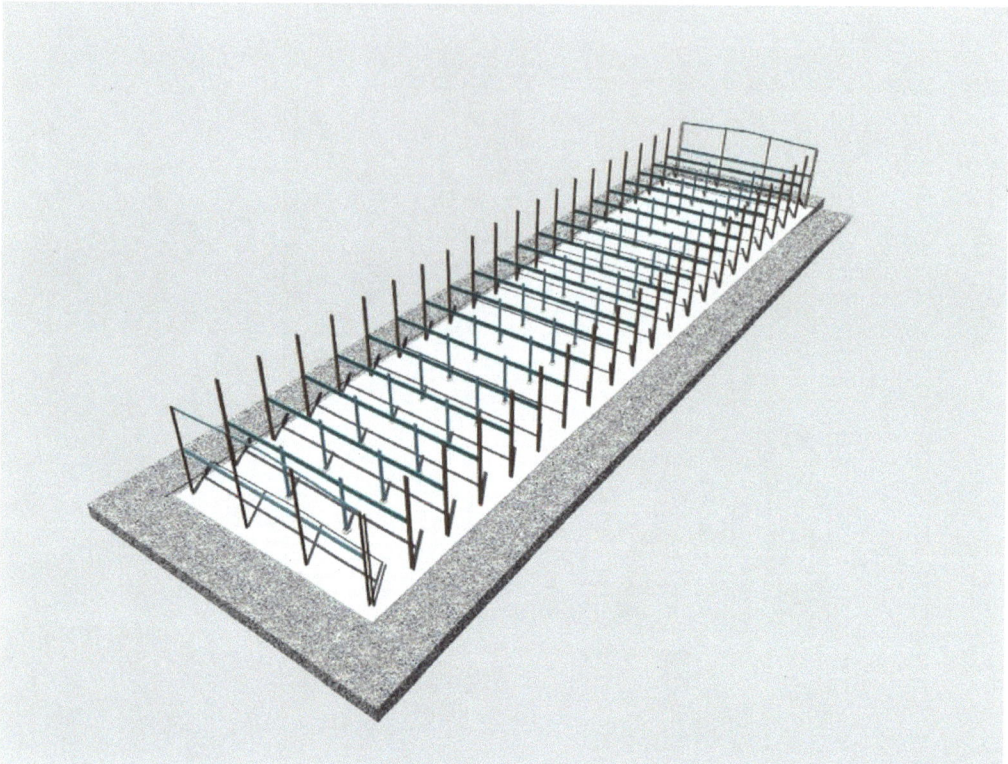

Steel beams which will be imbedded in the north (foreground) and south (distant) walls were placed at each end of the structure.

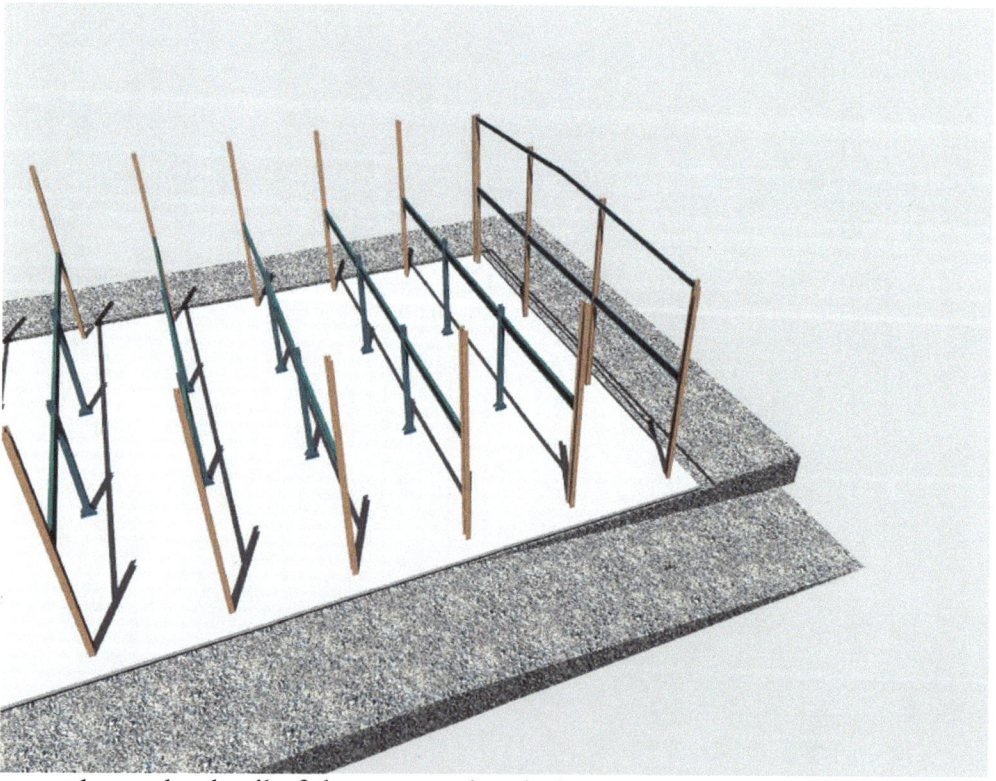

This image shows the detail of the structural end pieces for the end walls. This is the south end of the building.

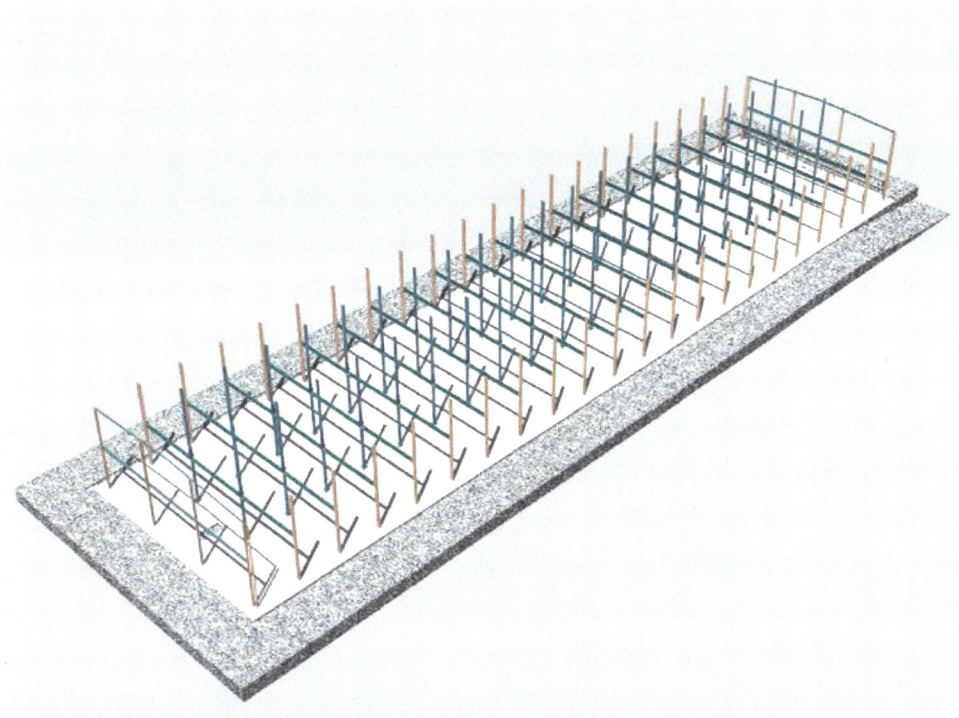

This image shows the vertical columns (in blue) with the second section for the second floor added, increasing the height of the columns.

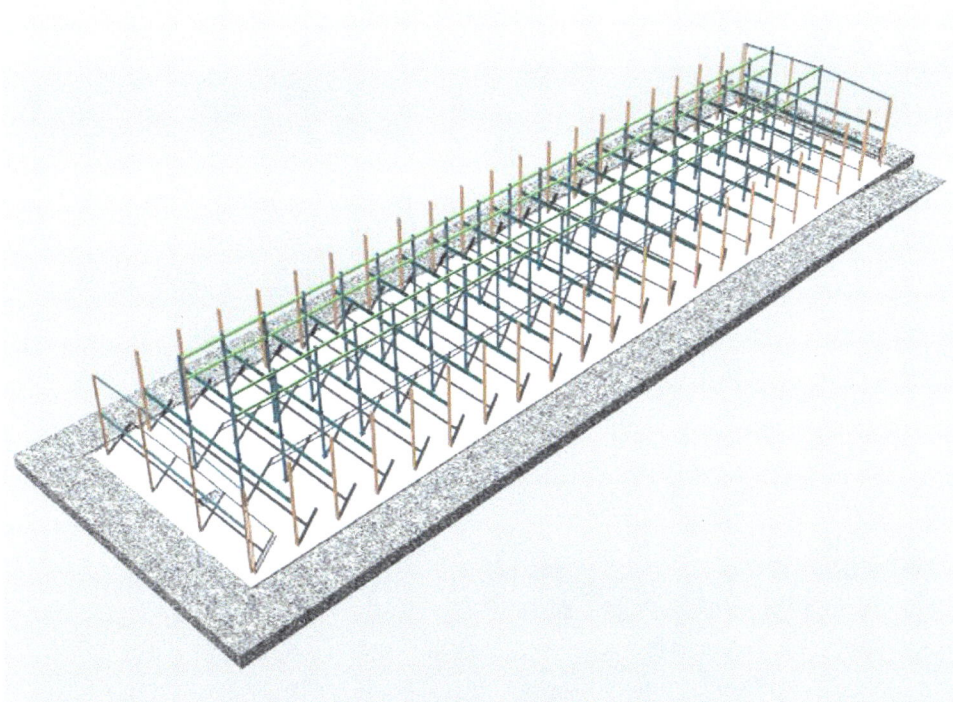

This view shows the 10" X 8" beams (light green) that will support the brick walls of the clerestory.

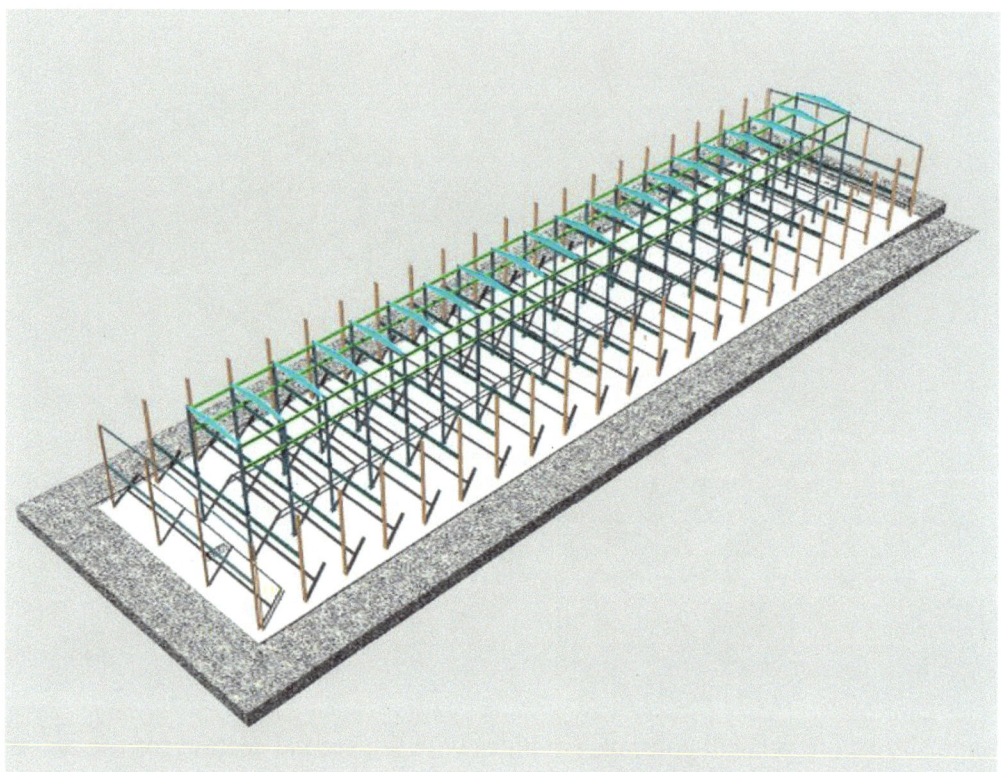

Across the central columns at the top of the structure are the triangular supports (turquoise) for the roof above the clerestory.

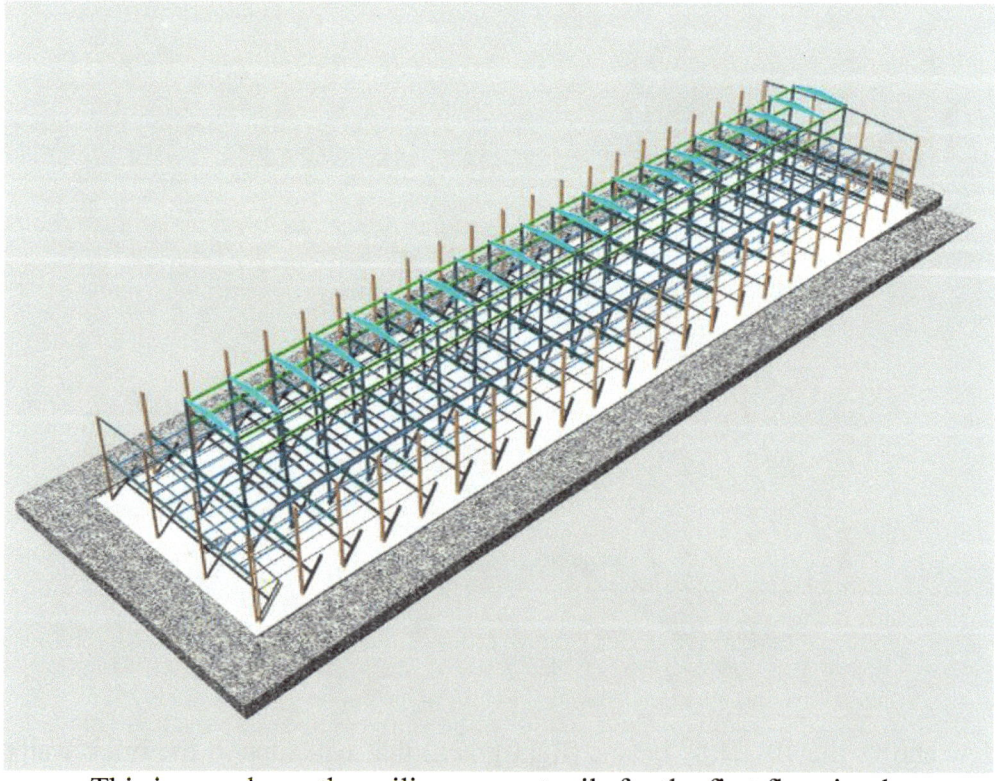

This image shows the ceiling support rails for the first floor in place.

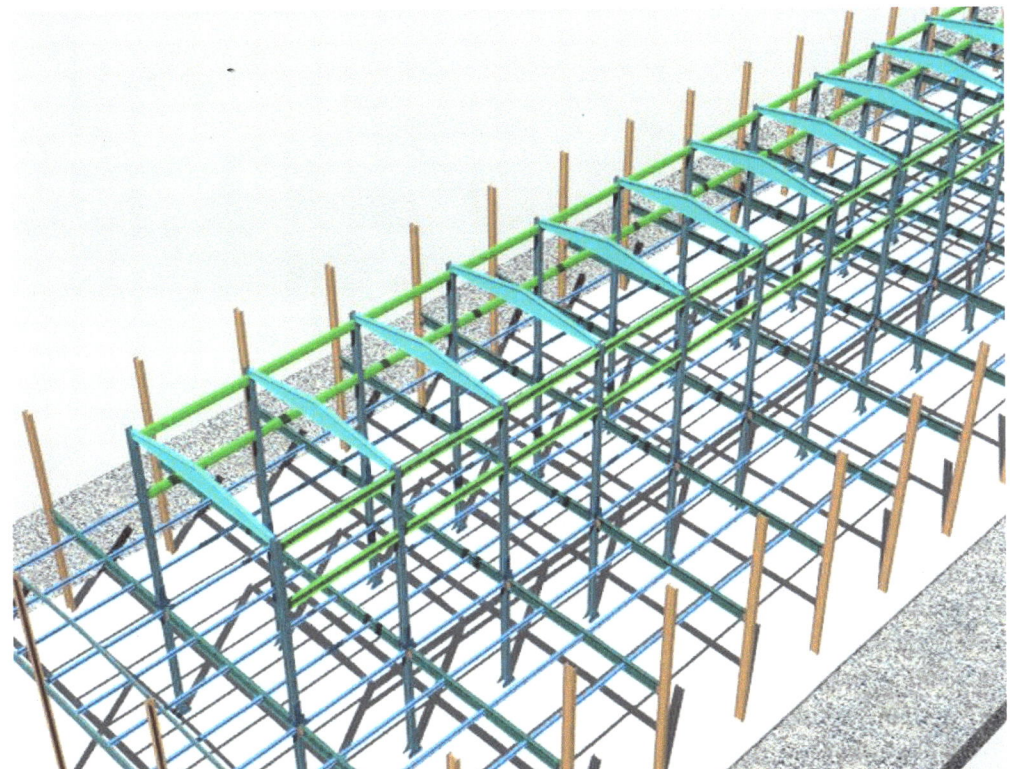

This image shows in better detail the first floor ceiling support rails (light blue). These supports were spaced seventy-two inches apart.

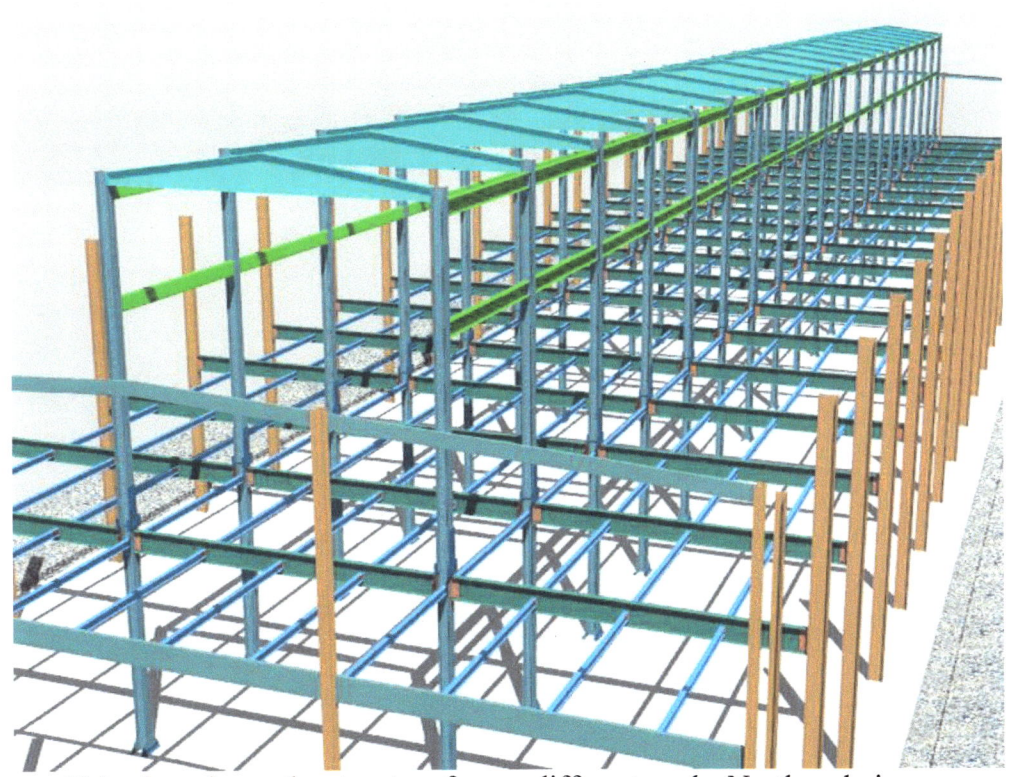

This view shows the structure from a different angle. North end view.

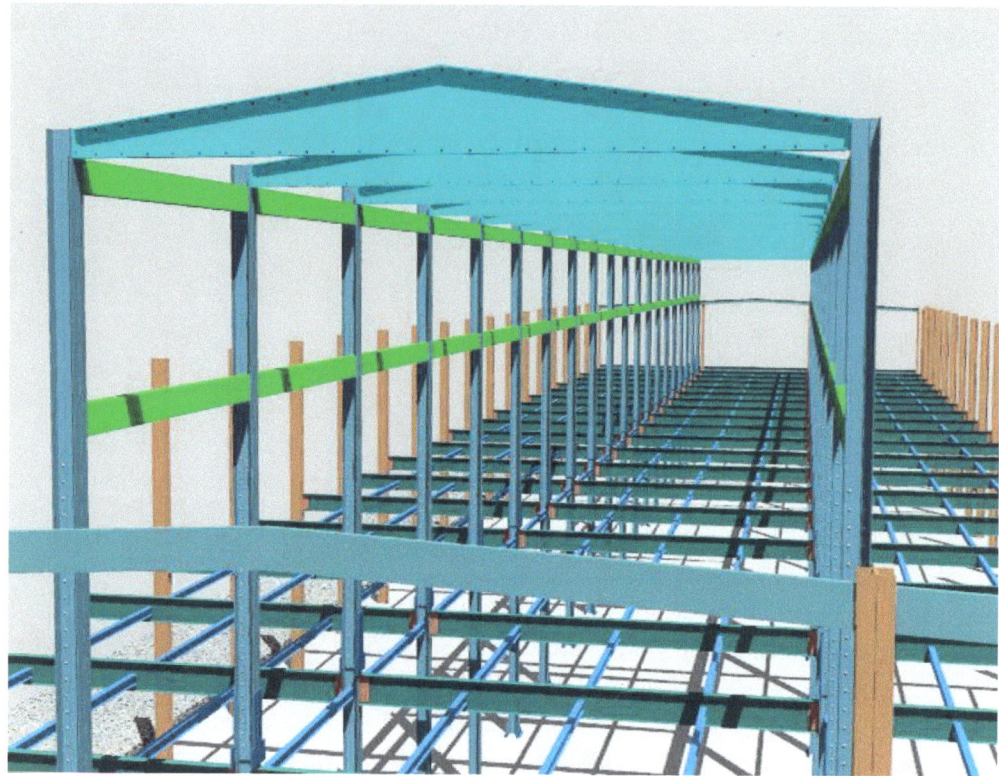

This view shows the construction detail of the upper clerestory structure.

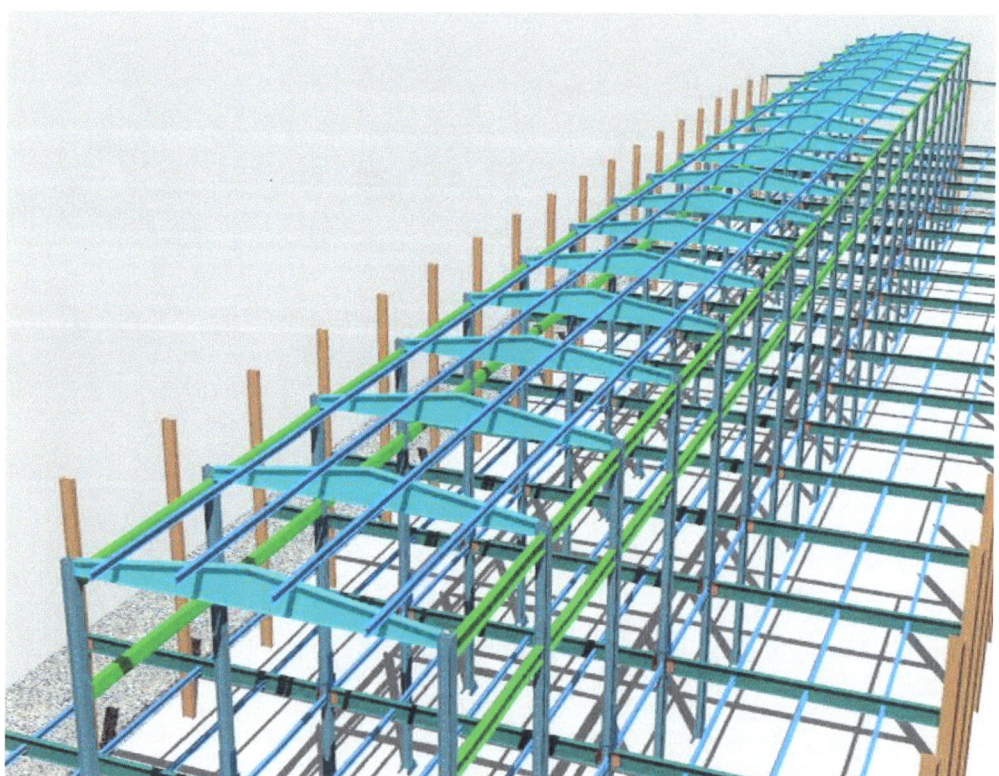

Across the triangular upper supports are placed the ceiling support rails for the clerestory roof arches.

CONSTRUCTION DETAIL
First Floor

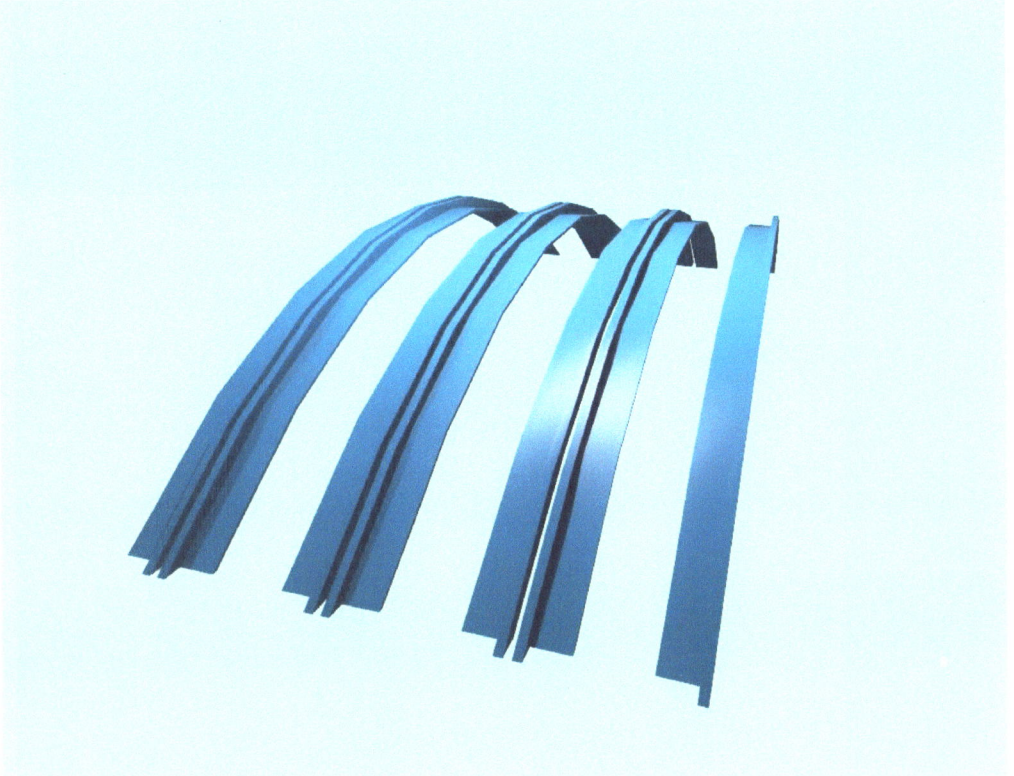

Steel arches used for the ceilings were made to hold fire brick between them.

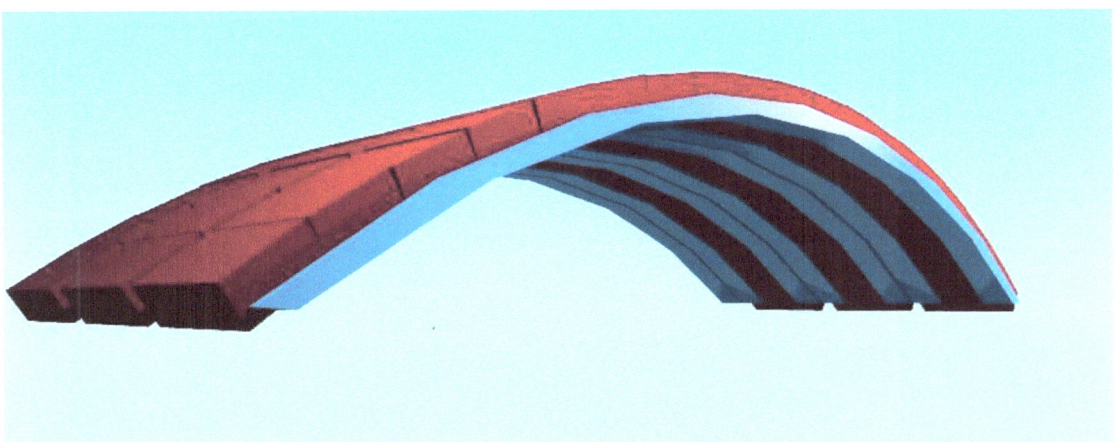

The fire bricks were placed between the arches in rows which became the fire proof ceilings used throughout the structure.

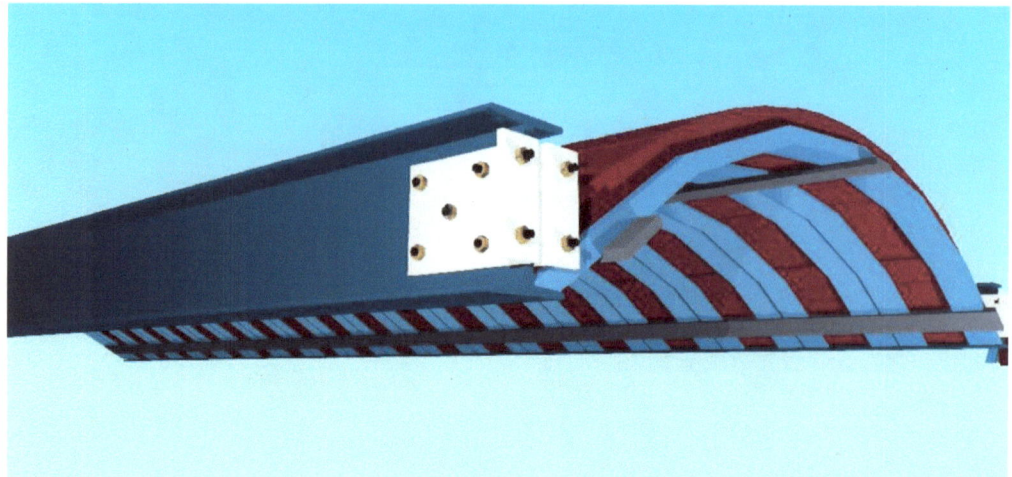

The steel arches were placed between six inch "I" beams spaced approximately seventy-two inches apart.

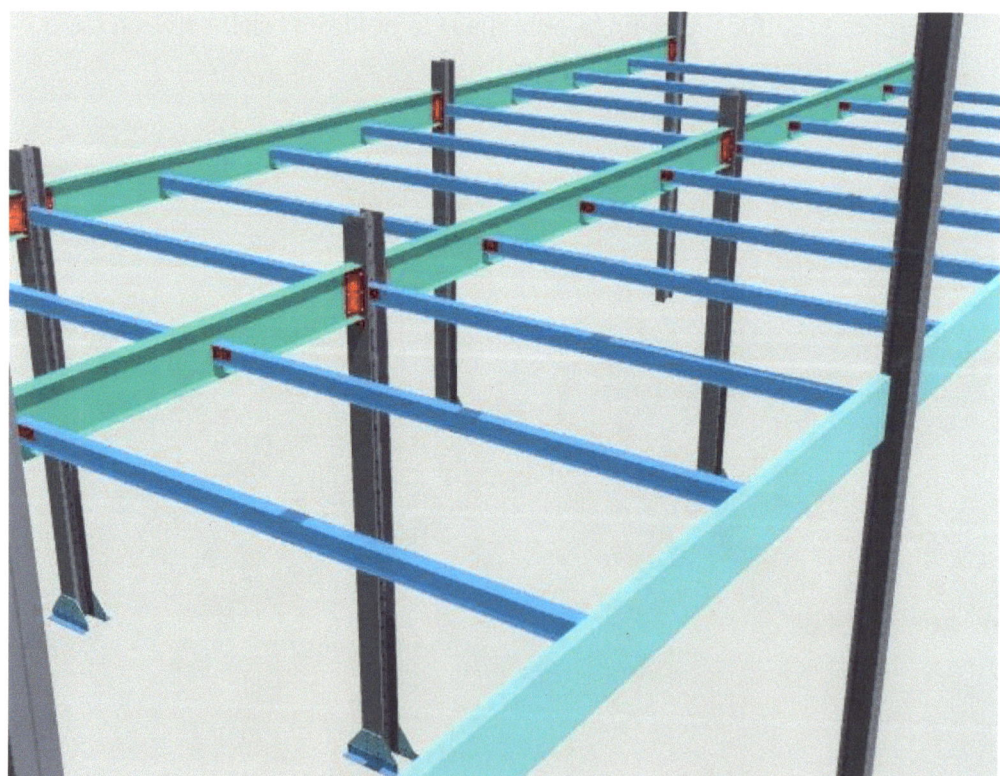

This view shows the six inch "I" beams across the main eighteen-inch structural support beams. These six inch beams are spaced approximately seventy-two inches apart. The structure shown is at the south end of the building connecting to the beam that is imbedded into the south wall. The beam connecting flanges are shown in red.

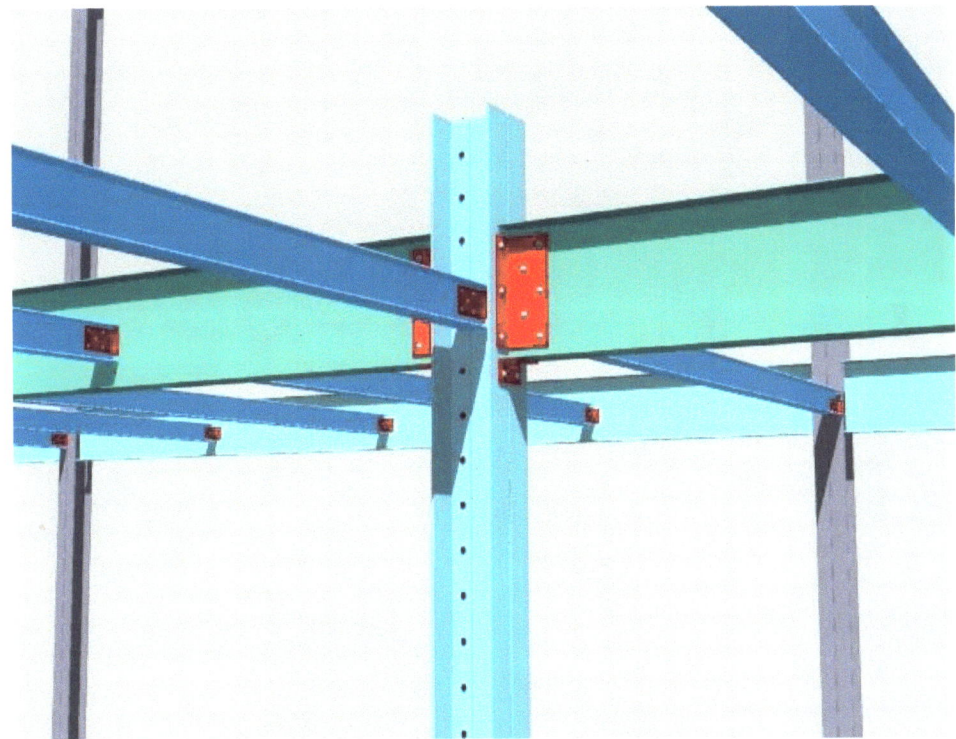

This view shows how the six inch and eighteen-inch width beams are connected by flanges riveted to the vertical columns.

This view shows how the eighteen-inch width steel beam is connected to the outside vertical support column. Brick walls with windows were constructed between each of the outside vertical support columns.

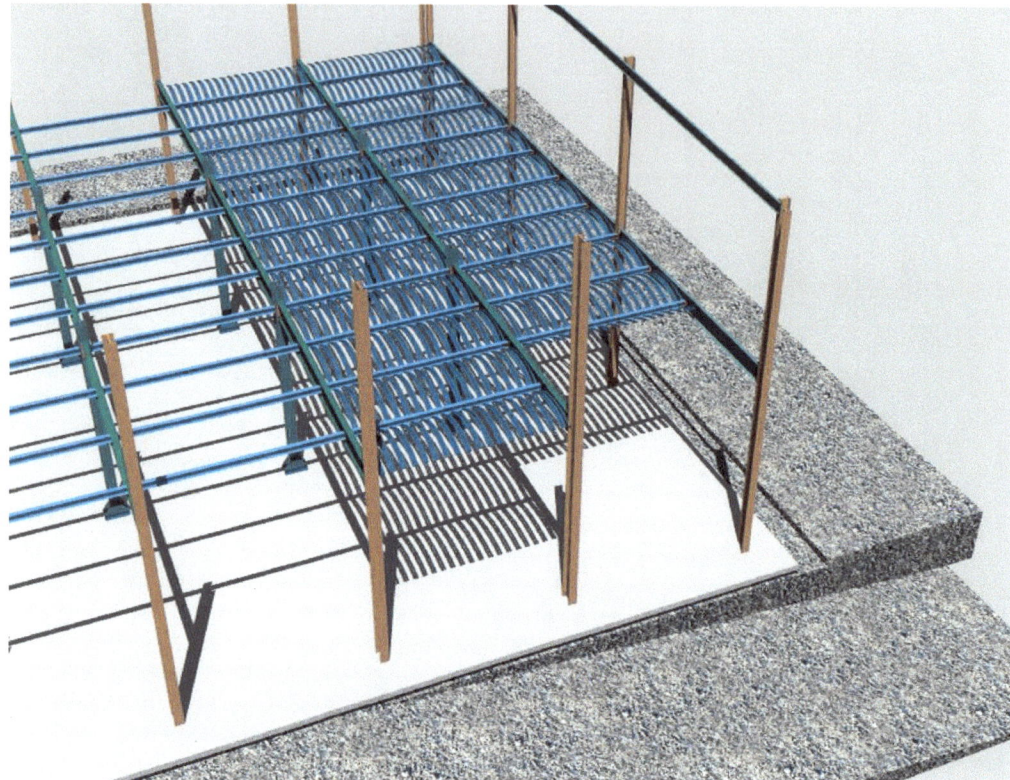

Arched steel ribs designed to hold the fire brick that was used in the ceilings of the structure are shown in place at the south end of the building.

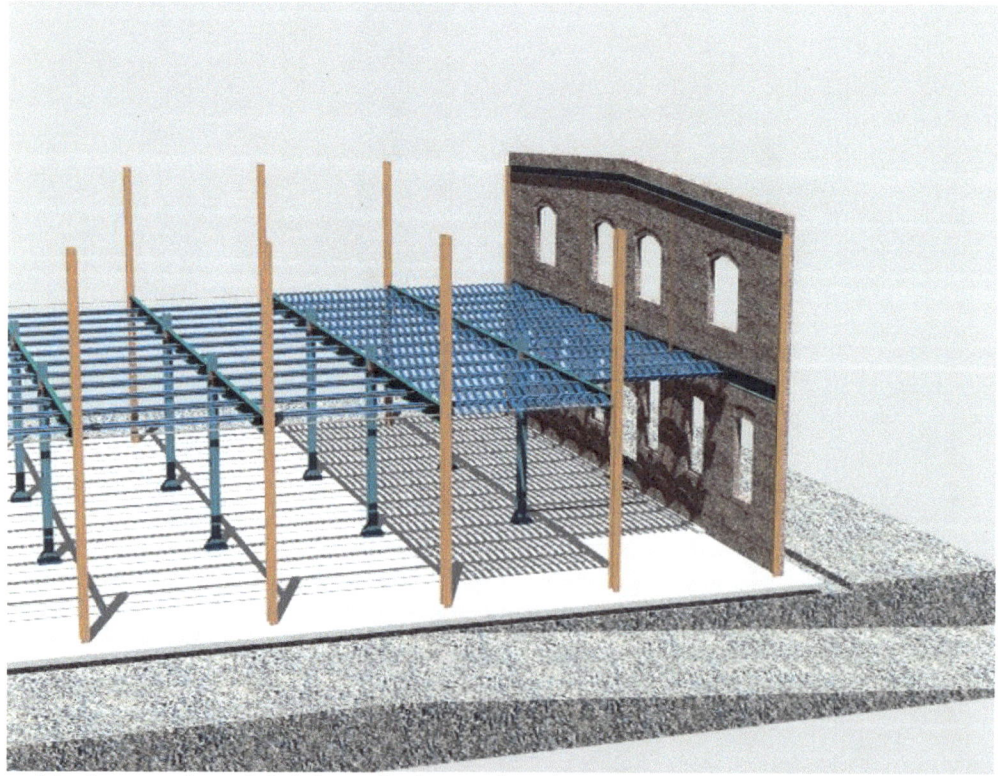

Image with the south end brick wall shown in place. Note the inclined ramp used to get material to the height of the first floor.

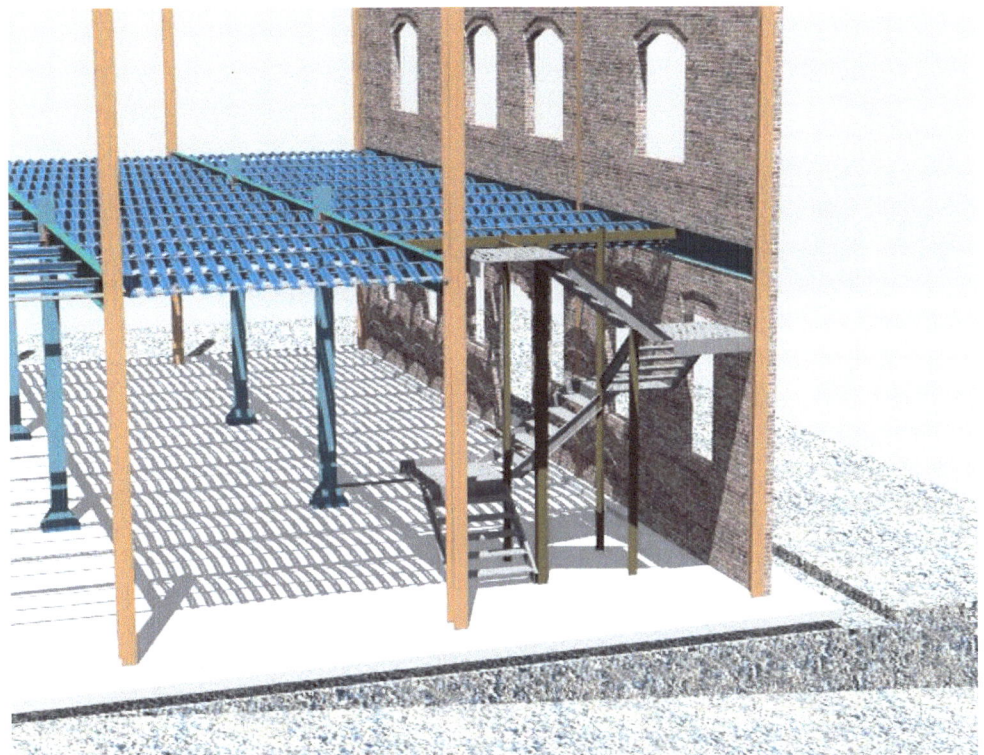

Stairs to the second floor are shown without the stairwell walls in place.

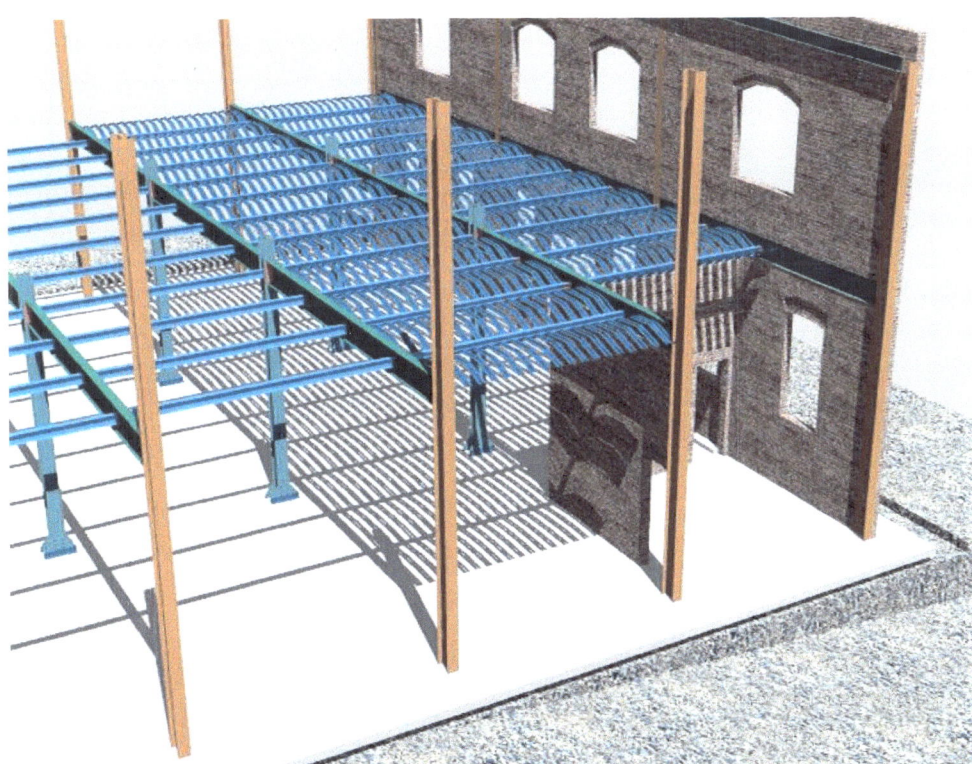

Stairwell-1 walls originally had two openings to enter the first floor area. One opening was next to the south wall and one near the west wall (not show in this image). When a large compressor was added at a later time, the opening near the west wall was bricked in.

Image shows Stairwell-1. The entrance to this stairwell was through a door located in the west wall (not shown). Directly in front of the entrance door were the steps to the second floor. A passage led along the west and south walls to the first floor entrance. Note-in this image the opening near the west wall is shown bricked in.

Steps for Stairwell-2 located near mid center of the west wall (not shown).

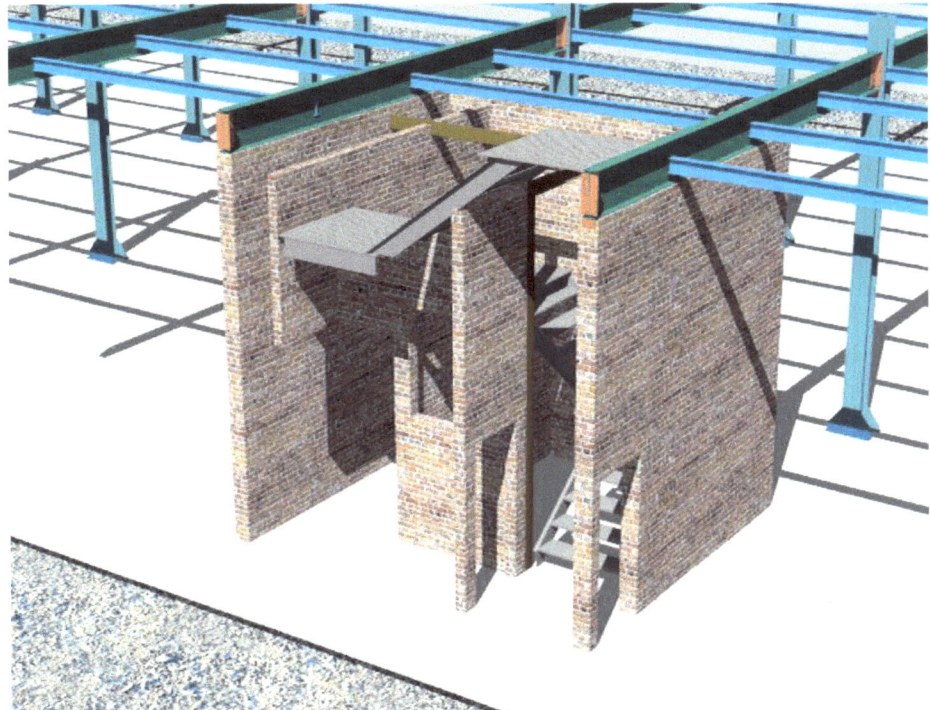

Stairwell-2 with staircase walls in place. From the entrance door in the west wall (not shown) directly ahead were the steps to the second floor. To the right was the entrance to the first floor, to the left was a passage and small storage area next to the left wall. An opening was made in the inner staircase wall to allow light into the stairwell from a window in the west wall.

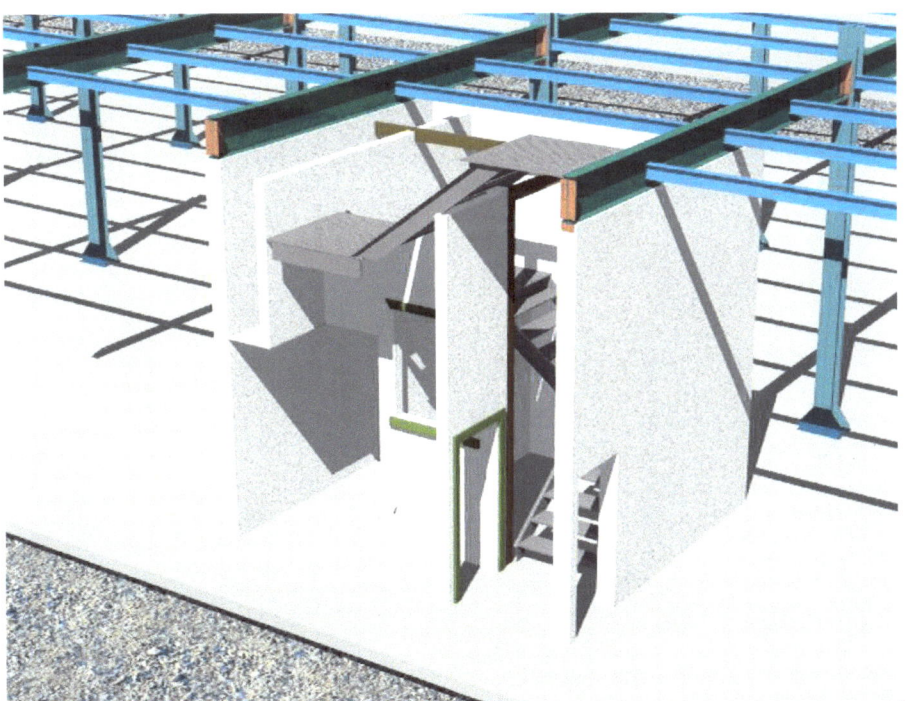

This image of stairwell-2 shows it with stucco applied so the door and wall openings with moldings can be more clearly seen.

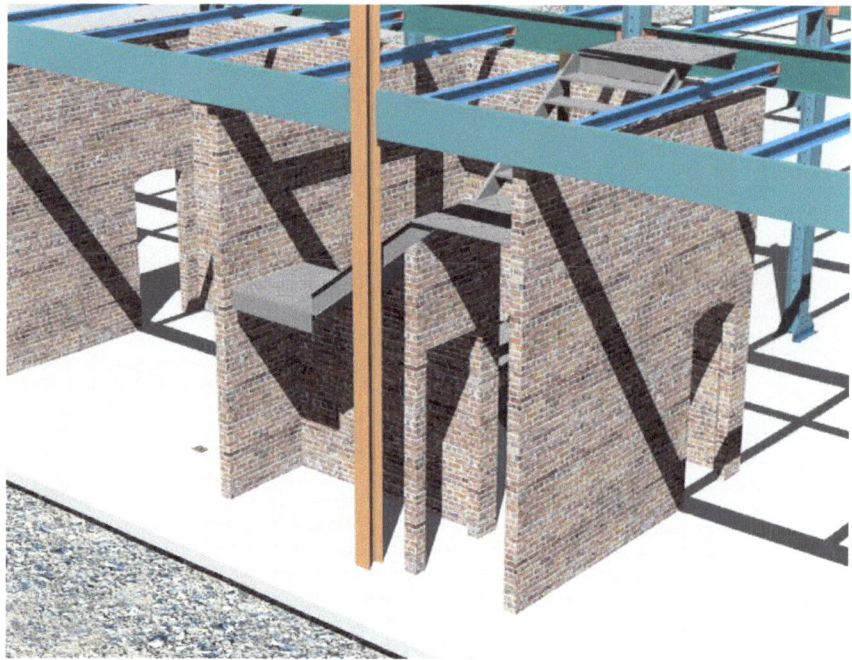

This image shows Stairwell-3 at the north end of the building. North wall is not shown in this image. Entering from a door in the north wall, to the immediate left was a small office. Going directly ahead led to a Staircase 3 on the left which went to the second floor. Opposite the staircase was an opening into the first floor area as seen on the right wall of this image.

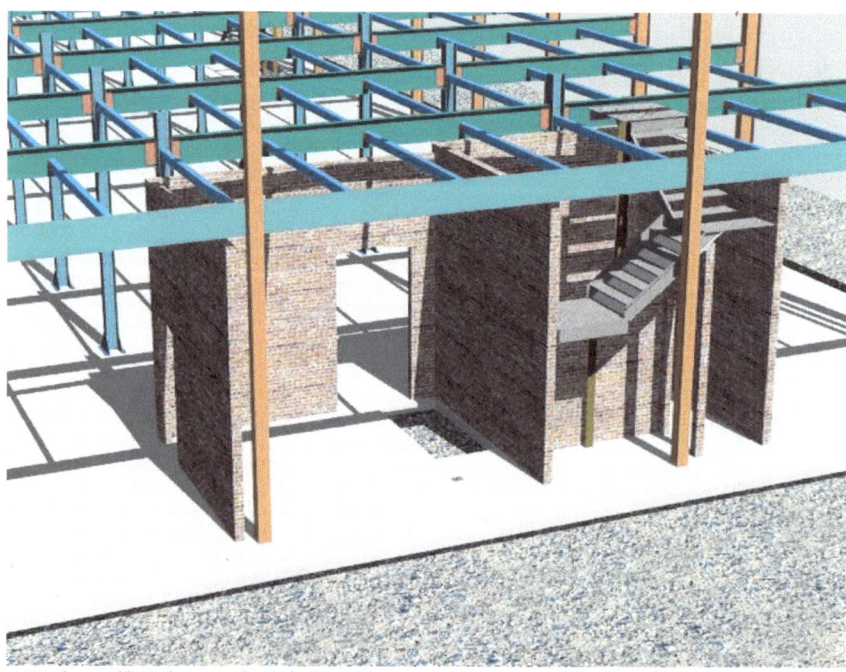

In this image Stairwell-3 is on the right. Center area was for receiving materials. A lift to the second floor was located here (note hole in floor) but was removed at some later time. Entrance to this receiving area was through a large double door in the north wall (not shown), a rear wall opening to the first floor, and the opening in the wall on the left of this image.

The openings for Stairwell-3 and the lift mechanism are shown opening to the second floor.

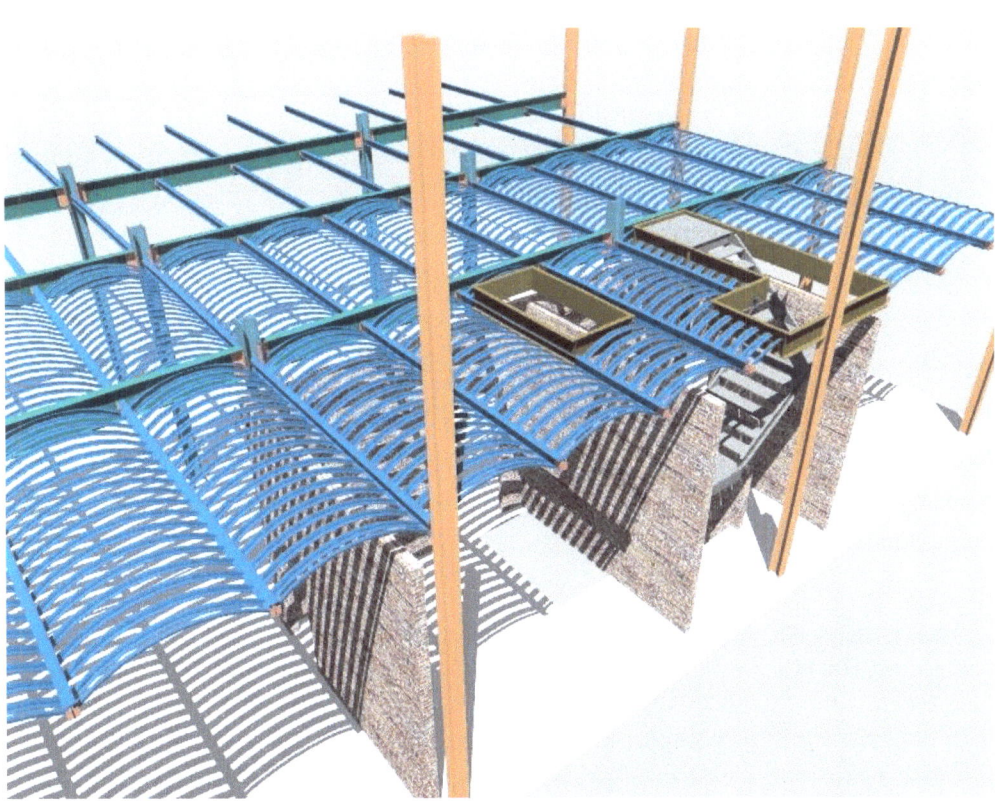

Another view of the openings for the lift and staircase to the second floor.

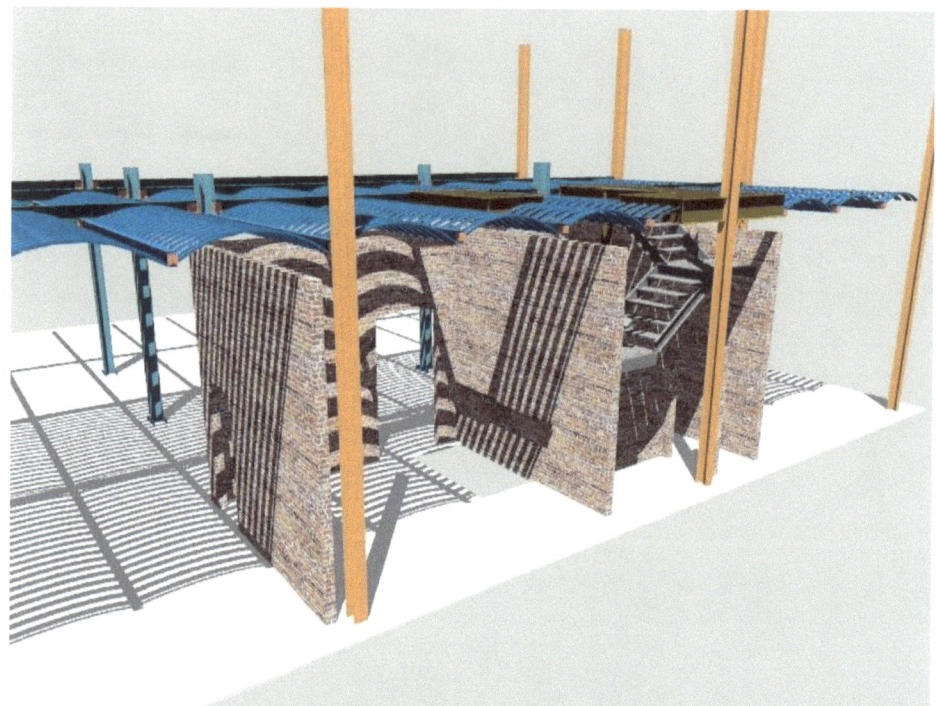

View of receiving area. In left wall of this image is the doorway into the receiving area (center) where the lift was located. Lift mechanism not shown.

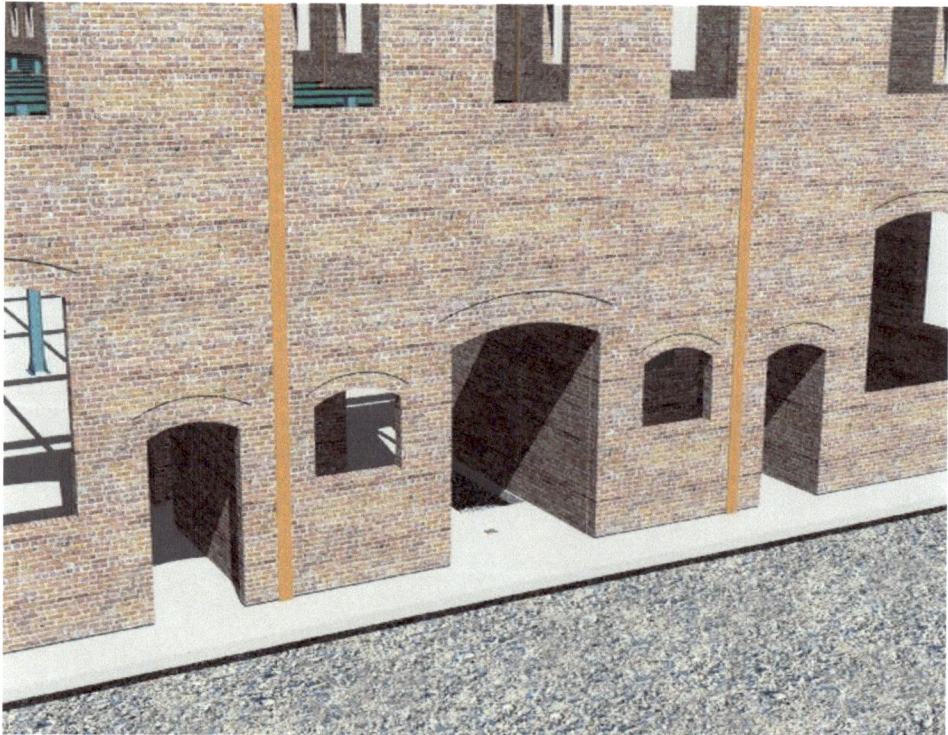

This view shows the rear or north wall of the building in place. The doorway opening on the right was the entrance to Stairwell-3, the small office on left with a small window, and an opening to the first floor. The center large opening was for receiving materials that could be lifted to the second floor, or moved into the first floor area. Receiving area also had a small window to the left of the large door opening. The doorway on the left opened to the first floor with an opening into the receiving area in the wall on the right of the doorway

4-FIRST FLOOR

INTERIOR VIEWS

The following illustrations show the first floor from the interior of the building. The stairwells will be shown as well as the general layout of the open space from the south to the north wall.

In some illustrations the plaster over the ceiling supports is shown, in other places the plaster is missing. Showing the plaster work completed would hide certain construction details to be explained.

There were forty main central vertical columns which formed the core of the building and distributed the weight of the structure to the ground. These columns were arranged twenty to a row.

These illustrations begin at the south wall and Stairwell-1 and proceed northerly to Stairwells-2 and 3. Stairwell-3 is located at the rear or north wall of the building.

This view looking southwesterly shows the south wall on the left, and Stairwell-1 in center. The wall in the background is the west wall of the building. In the foreground are some of the main vertical columns of the structure.

This view shows the entrance to the first floor from Stairwell-1, the window light illumines the entrance at the south wall. Originally there was also an entrance at the west wall, but it was bricked in when the boiler and compressor were added to the building. On the stairwell wall near the window on the right in this image, an unknown artist painted an image of the face of the former President of the United States, Ronald Reagan. It is believed to have been painted sometime during the 1990's, after the boiler/compressor were removed. Photo 6, 6A, 6B, 6C.

Double doors opened to the center of the first floor. Stairwell-1 is to the right.

The main central doors in the south wall had provision for narrow gage track, leaving an eight-inch space on either side of the central metal plate. This narrow gage track was for rolling carts or platforms to move material through the building. They were not normal sized railroad track. It appears track in this area was never put in, or it was removed when the boiler/compressor were added to the building. Photo 7.

In the area of the south wall and Stairwell-1 a large air compressor and boiler were added to the building after the original construction. From photographs it appears the original wooden flooring was removed and concrete poured to support the weight of the foundations supporting the equipment. The illustration above shows wooden flooring which may have been there prior to the equipment being added.

Interior view of Stairwell-1. Door directly ahead is the entrance into Stairwell-1 from the exterior of the building. The widow to the left of the door is to let light into the corridor leading to the first floor entrance way at the south wall.

This view is looking towards the south wall. Along the west wall, on the right of this image, can be seen the windows and two sets of large double doors opening to the first floor. One row of the main structural columns can be seen along the left edge of the image. Photo 8.

This image shows the entrance to the first floor from Stairwell-2, located along the west wall. The doorway entrance through the west wall is visible in the near center of the image through the first floor entrance from Stairwell-2.

This image shows the stair arrangement and the door and window which are within Stairwell-2. Brickwork of Stairwell-2 is not shown in this image.

Stairwell-2 had an interior wall with an opening in it (center of image) which let light from the window in the west wall enter the step area. Photo 9.

Light entering through the outer west wall is let into Stairwell-2 by an opening in the interior stairwell wall allowing light to enter the inner stairwell.

Interior of Stairwell-2. The door is the entrance to Stairwell-2 from the exterior of the building. To the left of the entrance door is the opening to the first floor. In the center of the image is the opening to let light into the staircase. On the right is a wall behind which is a storage area. Photo 10.

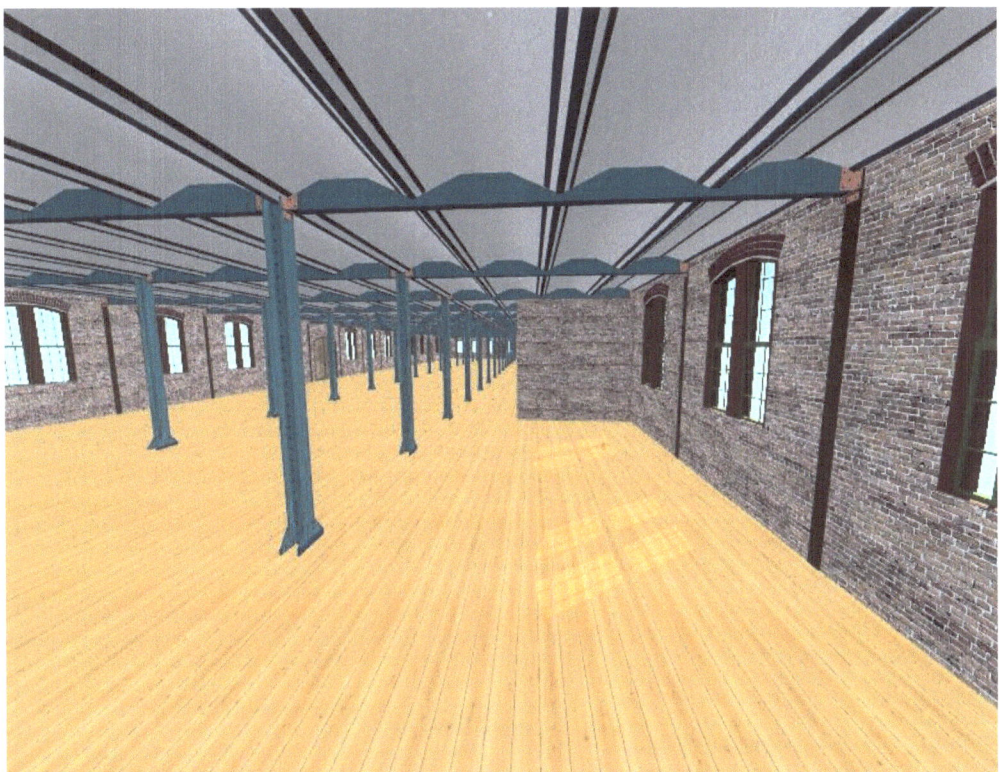

This view is looking towards Stairwell-2, from the north. Photo 11.

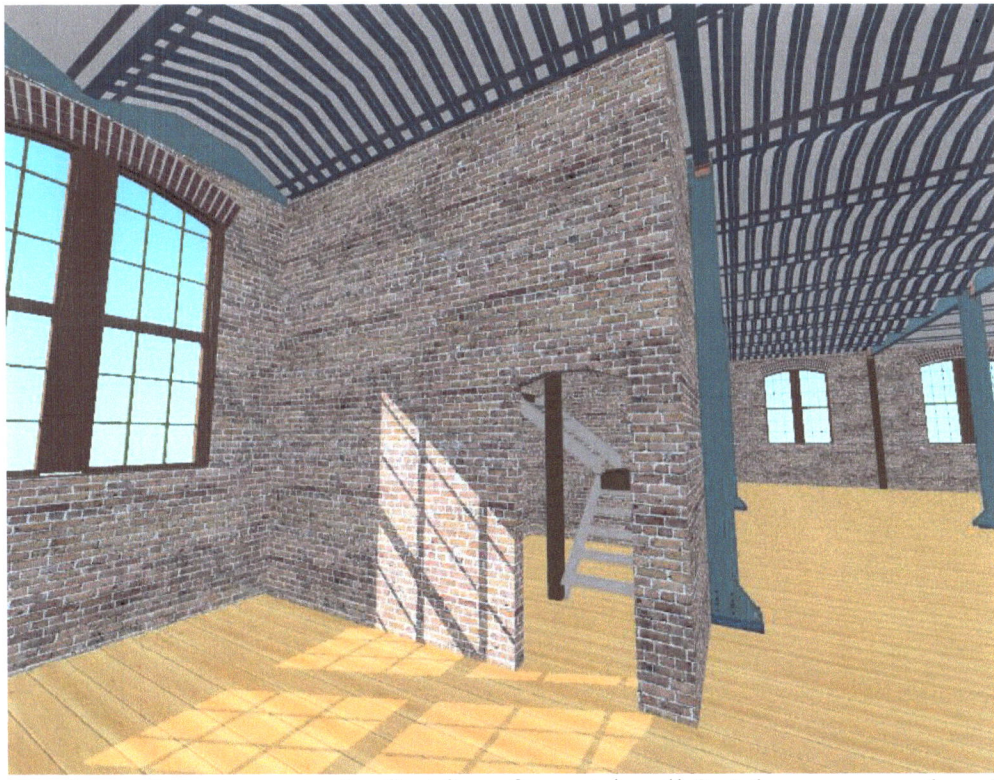

This view shows the entrance to the first floor from stairwell-3 at the rear or north end of the building. The west wall would be behind the viewer from this position. View is looking northeasterly. Photo 12.

This view shows the large opening into the first floor from the receiving area located at the north end of the building. There was one window in this area for admitting light.

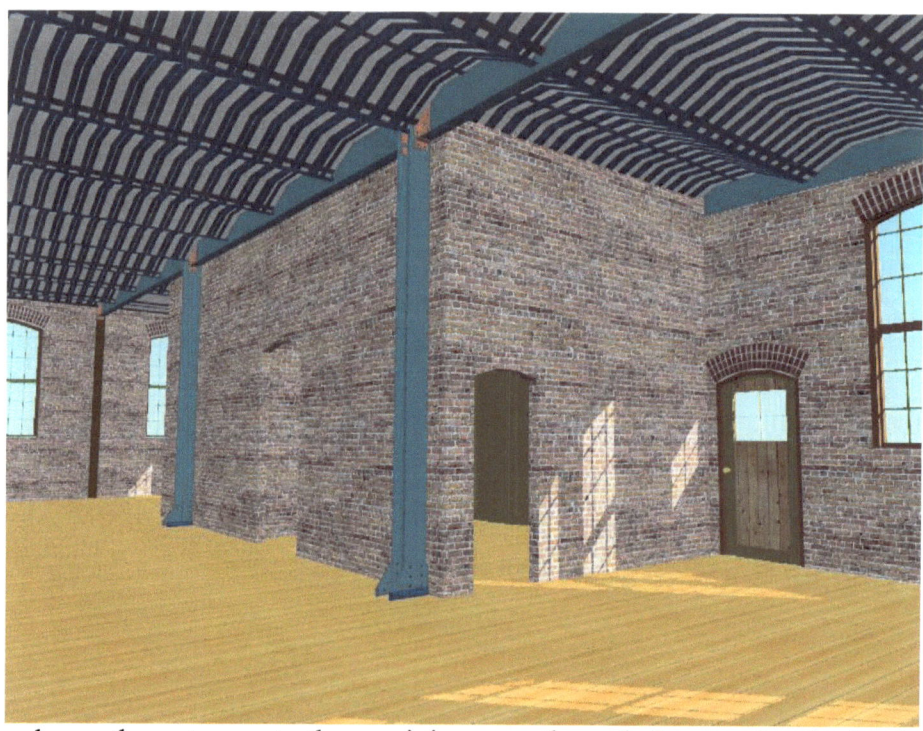

This view shows the entrance to the receiving area through the opening in the foreground. Near the widow on the right is the rear door which opens to the first floor from the exterior. Between the two vertical columns is the opening to the receiving area.

This view towards the rear, or north wall, gives an idea of the huge open space on the first floor. This image would be as if the viewer was in line with the first set of large double doors in the west and east walls near the south end of the building. Jutting from the west wall on the left is Stairwell-2. Directly ahead is the large opening into the receiving area with the central double doors in the north wall closed (green). Photos 13, 14.

The wooden flooring of the first floor when photographed in 2003 showed no signs of machinery having been bolted to the flooring. There were signs of flooring having been replaced, especially near the large doors in the east and west walls.

If work was performed on this floor there would have been signs of machinery marks, air pressure piping, or other signs of machinery having been used. Photos 15, 16, 17.

A newspaper article appearing in a 2006 edition of the "Morning Call", an Allentown, Pennsylvania newspaper, prior to this building being torn down, confirmed this building was used as a warehouse for parts storage based on a 1917 Lehigh Valley Railroad track map.

5-SECOND FLOOR

CONSTRUCTION & INTERIOR VIEWS

The second floor is shown separately from the first floor in these illustrations in order to make the parts of the second floor more distinctive and clear to the viewer.

These illustrations of the second floor show construction detail and how the floor space appeared when the structure was originally constructed.

The second floor had a distinct set of narrow gage rails running down the middle of the building. These tracks were used with movable wheeled platforms to move parts to different locations on the second floor.

Some type of mechanical lift, not present when this study was done, was used to lift loads from the receiving area on the first floor up to the receiving area on the second floor.

On the second floor a room at the north wall was used as an office, it being a part of the structure along the north wall including Stairwell-3.

The north wall had a different window arrangement and used some different style windows than those used in the south wall.

Fire walls were used on the second floor but do not appear to have been used on the first floor. Several sections between firewalls were subdivided on the second floor at a later time and are not shown in these illustrations.

The second floor had a clerestory running almost the entire length of the building. The main purpose of the clerestory was to admit as much light as possible into the second floor. Large multi-paned windows were located in every-other wall section along the clerestory of the building.

Though electrical lighting was just beginning to be used at the time of construction there was no evidence it was used in this building, as there were no electrical panels or wiring found to indicate the building had been wired and no lighting fixtures were found.

Electrical wiring standoffs appeared at various locations on the exterior of this building at the second floor level, these standoffs may have been added at a later time to hold wiring going to other buildings.

The beginning images of the second floor do not show structural parts such as the cross beams, ceilings and the clerestory structure for viewer clarity.

The center superstructure which supported the clerestory walls is shown.

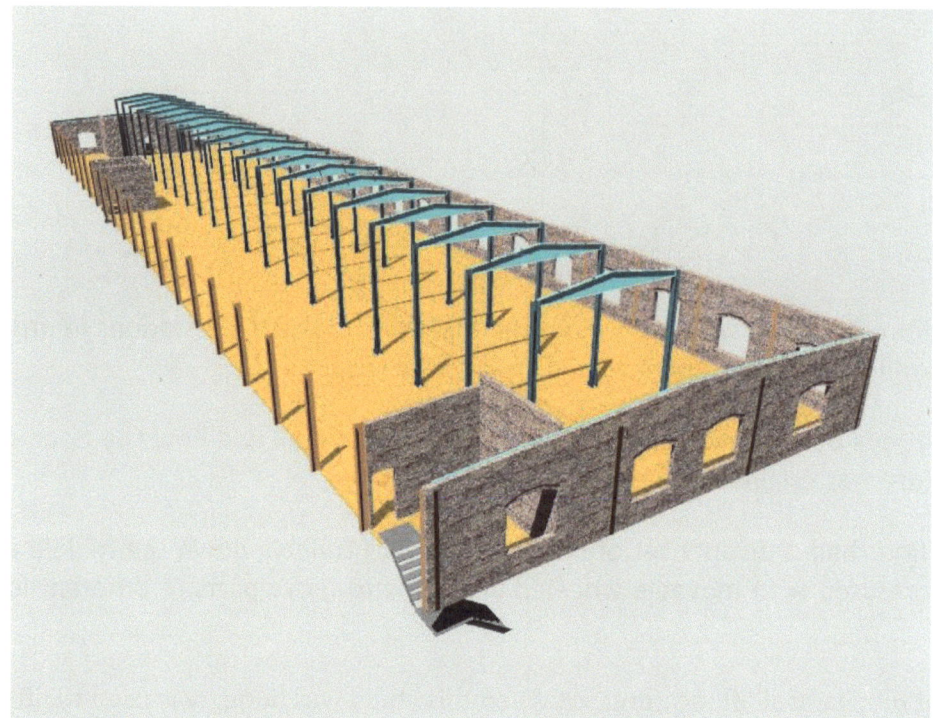

This view from the south wall on the second floor shows Stairwell-1 and the opening from the stairwell to the second floor. In the distance is seen Stairwell-2 and at the rear is the north wall and Stairwell-3. The second floor west wall is not shown in these images for clarity.

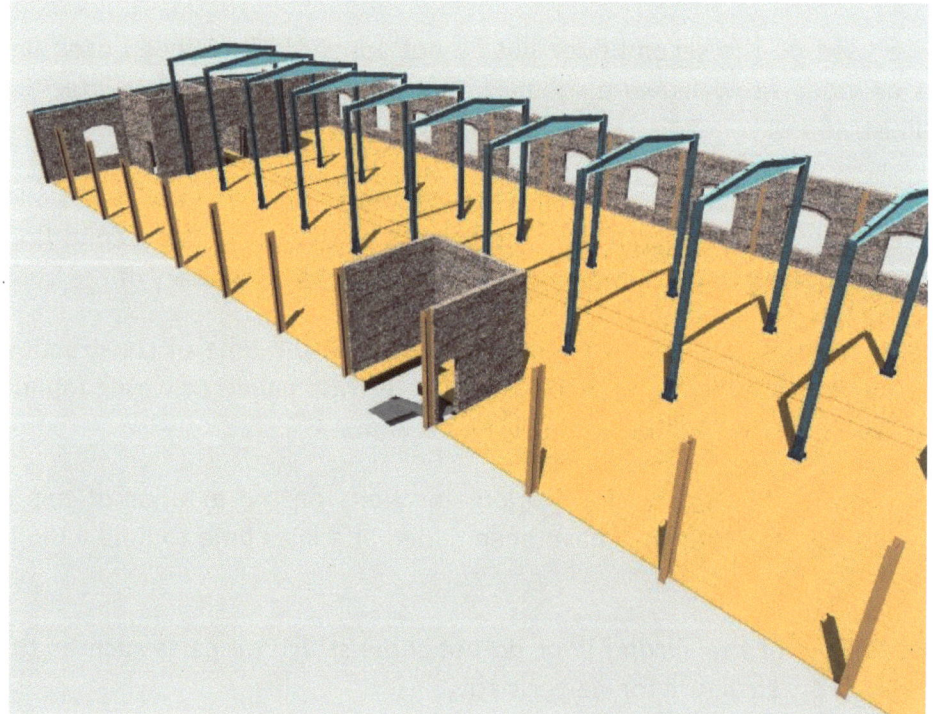

In the center of this image is Stairwell-2. The opening from this stairwell in the center of this image opened to the second floor.

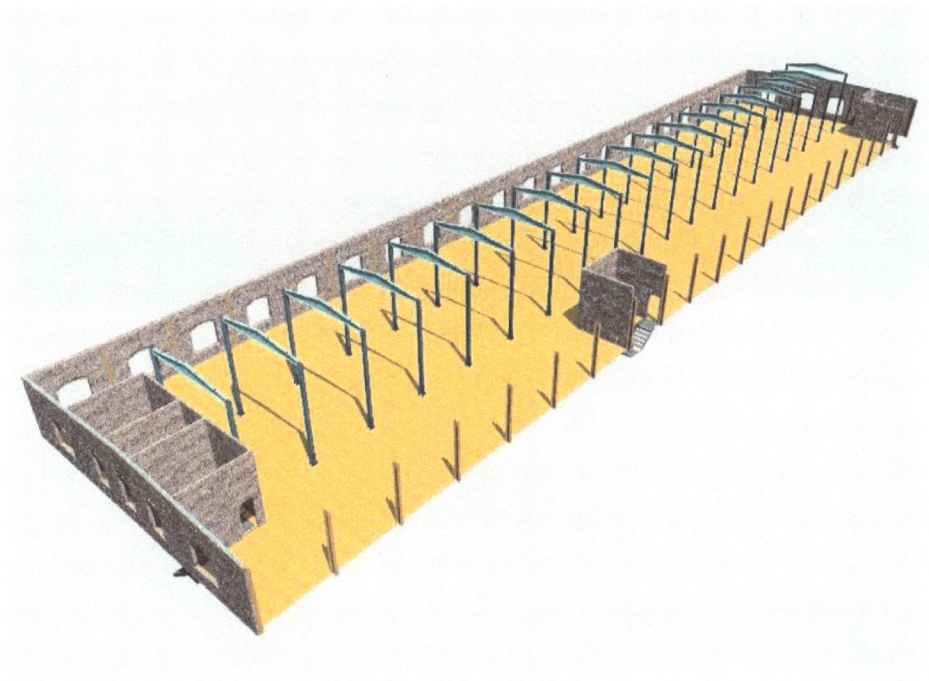

This image shows the north end of the building on the second floor. The rear entrance to the second floor is visible in the wall of Stairwell-3 on the left of this image. In the center of the floor is Stairwell-2, and at the top right is Stairwell-1.

This image shows the superstructure, ceiling structures, and support beams on the south end of the building. Entrance to this floor is visible in Stairwell-1 wall.

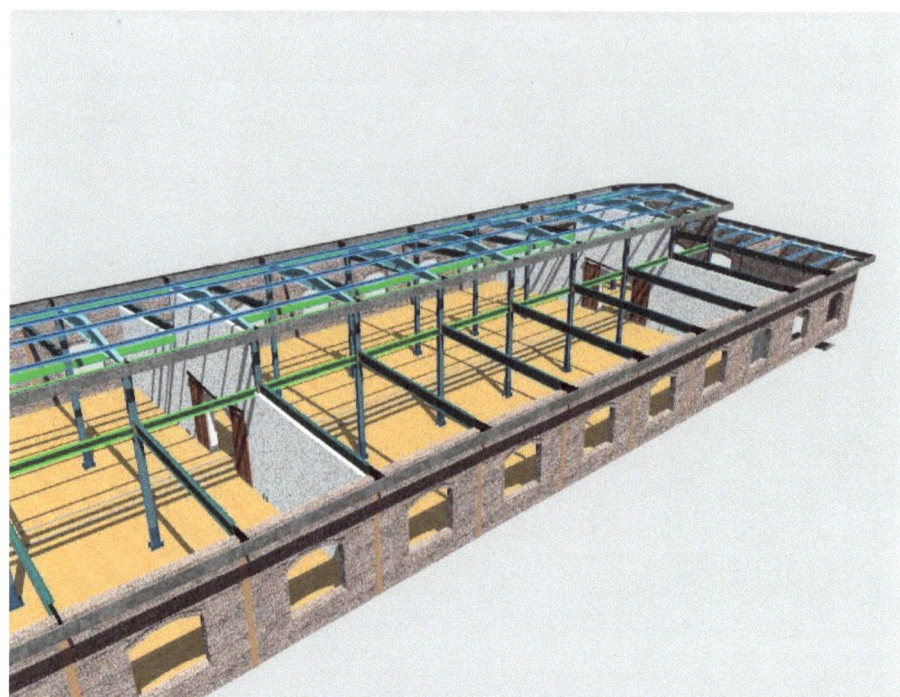

The second floor was divided into four separate spaces to prevent the spread of fire. Each fire wall had double fire doors which closed by gravity. The fire walls are shown covered with white stucco for clarity. The west wall is shown in these images.

Angular top view showing the four fire walls in place. The fire doors were placed on the fire walls facing the center of the building.

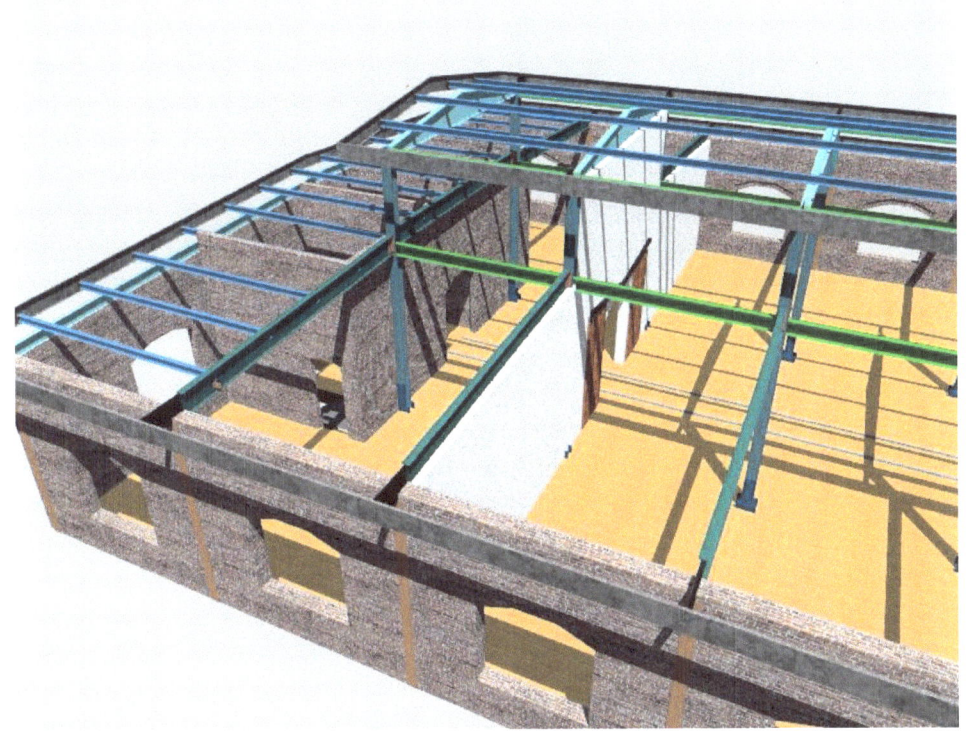

This image shows the fire wall at the north end of the building. Note the fire door arrangement.

This view shows the fire wall and Stairwell-3 (on lower right), lift opening (center), and office area on left. Rear or north wall appears at the bottom. East wall not shown in this image.

Clerestory brickwork and ceiling structures are shown in this image. The steel ceiling arches and supports that cover the entire second floor are shown. Fire brick was placed in these arches.

The fire brick was then coated with a layer of plaster making the ceilings very resistant to fire. This same technique was applied to the top roof above the clerestory.

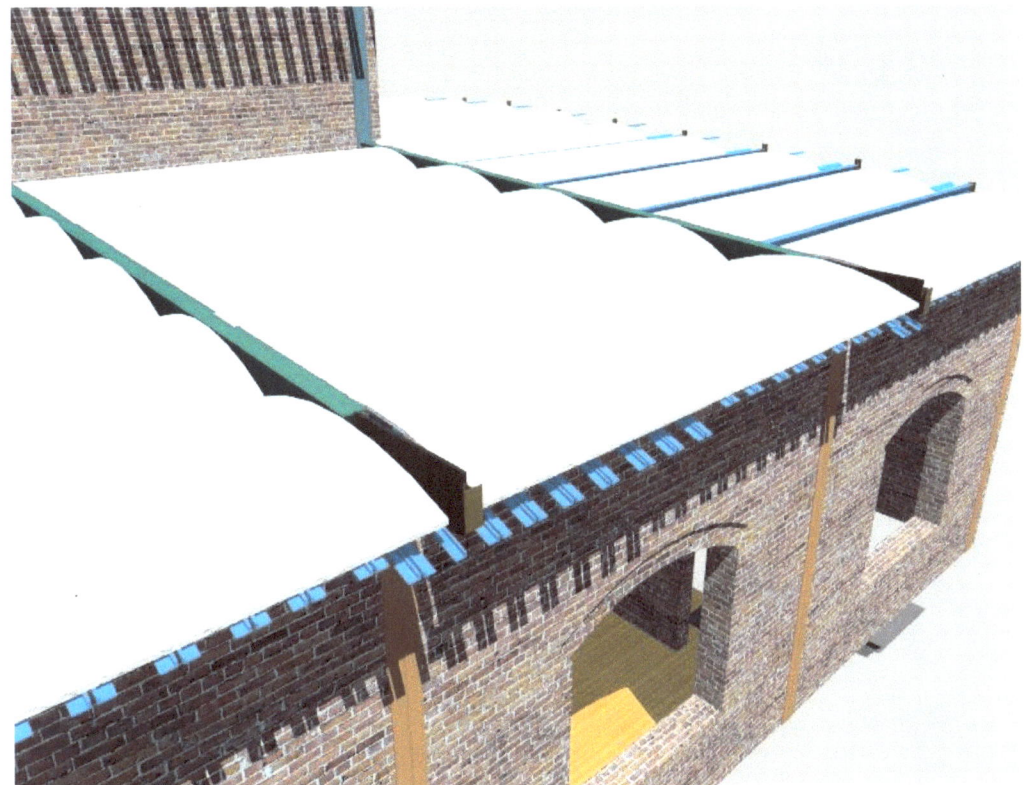

At the end of each steel eighteen-inch crossbeam was added a supporting device to hold the rain gutters and outside finishing steel banding.

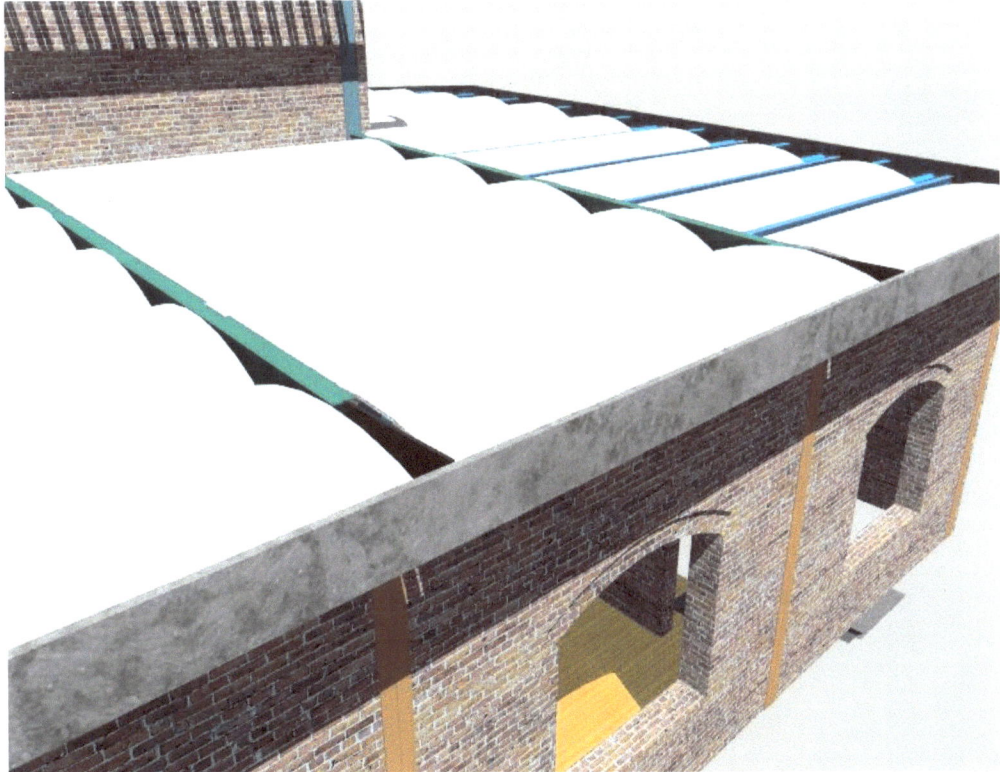

Here the outside finishing steel band is shown in place. The roofing was then put on the structure.

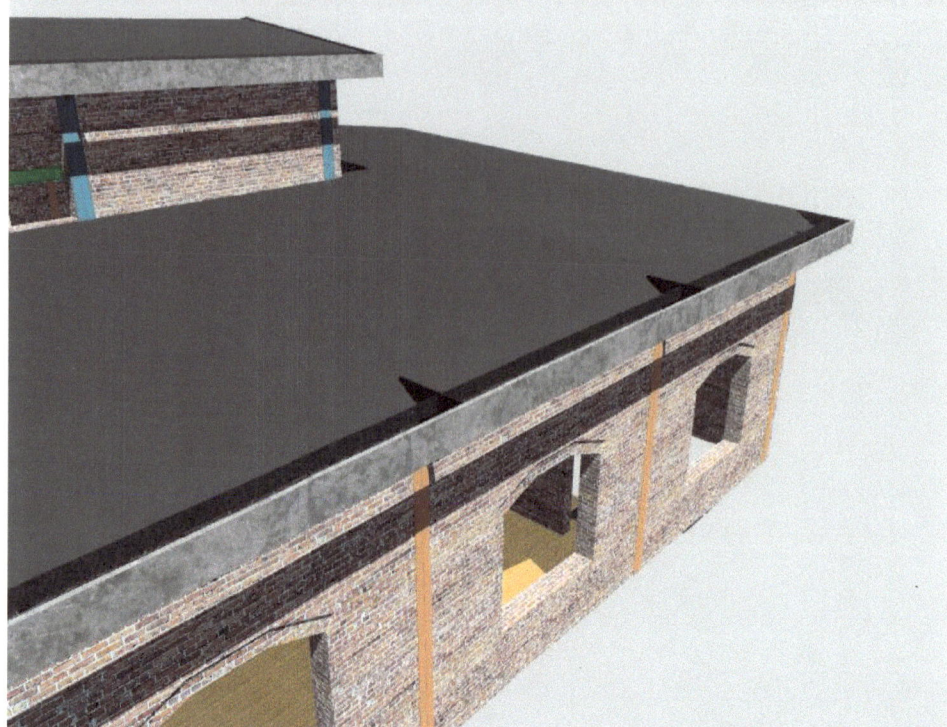

This view shows the asphalt roofing in place with the steel banding on the clerestory and second floor level.

This image shows the stairs from the first to the second floor located in Stairwell-1. Photo 18.

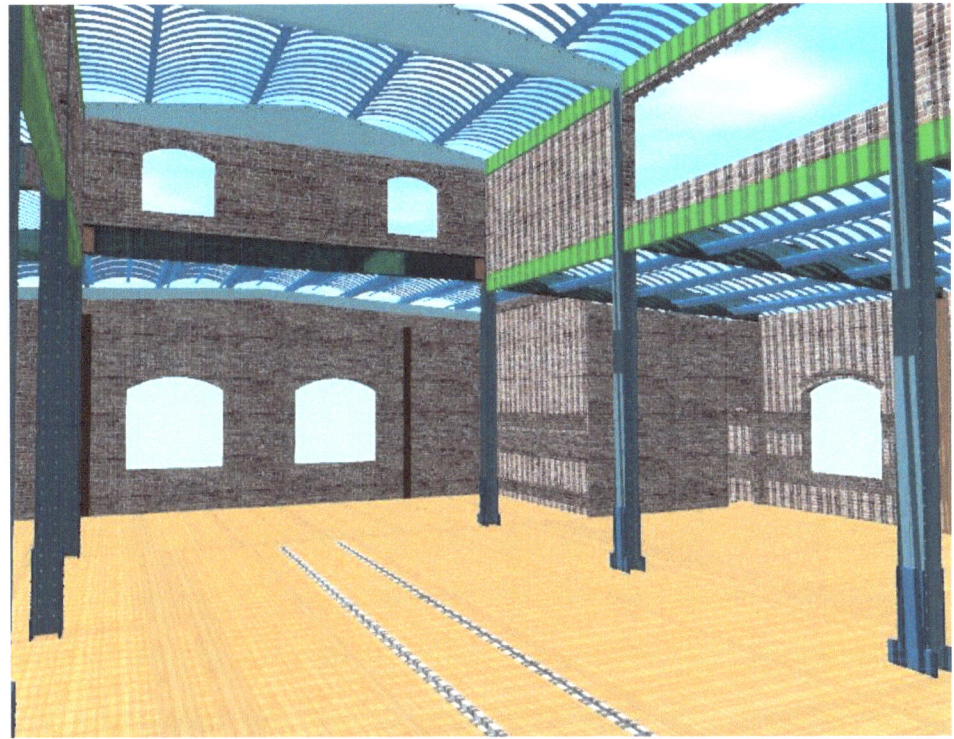

This view, looking towards Stairwell-1 in the right corner, shows the second floor prior to the windows and roofing being completed. Note the central cart tracks and the plates on each column to support the added column length for the second floor. Photos 19, 20.

View looking south. Stairwell-2 juts out from the west wall located on the right. In this view all the windows and clerestory windows are shown.

This view shows the entrance to the second floor from Stairwell-3. In the center is the large opening for materials lifted to the second floor and the tracks running down the center of the floor.

This view shows the rear receiving area opening, lift opening in the floor, and tracks. The clerestory with some of its windows is also shown.

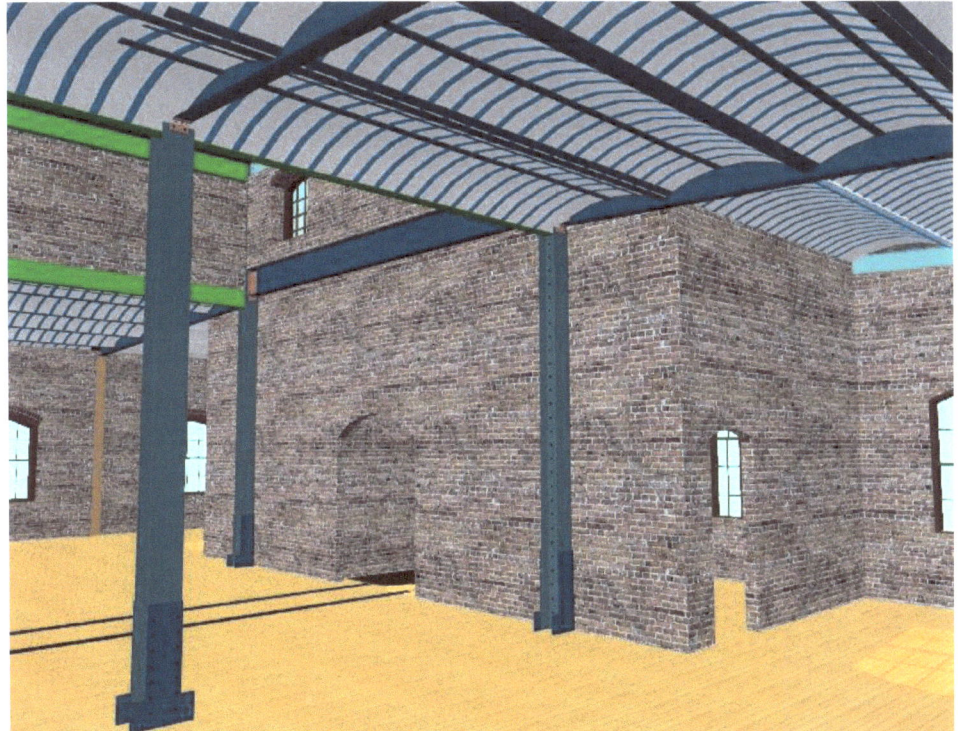

This view shows the entrance to the office area on the second floor in the wall on the right. This is at the north end of the building.

View towards the south end of the building. Stairwell-2 is on the right. Cart tracks run down the center of the floor.

View towards the north end showing the fire walls in place. Doors were on angled sliding mechanisms which kept the doors closed automatically when not purposely held open. Photo 21.

View looking north through a firewall opening, with the cart tracks in the center. The two interior firewalls in the south end of the building had the fire door mechanisms opposite the openings.

This view shows a fire wall in the background, the west wall on the right, and the ceiling structure. This was the typical construction used throughout the building. Photo 22.

This completes the study of this interesting building and the techniques used in its construction.

This structure was one of the earliest steel skeleton buildings used for industrial purposes in Pennsylvania. It was built at a time when steel core buildings were the new trend of the future and the days of using large wooden timbers for the construction of mills and industrial buildings was past.

This building was the last structure left standing in what was once a large railroad terminal called the Packerton Yards of the Lehigh Valley Railroad.

This building was completely demolished in 2007, leaving no trace of its former existence.

Lou Robertella-2016.

6-SUPPORTING PHOTOGRAPHS

The following photographs were taken during the original visit to this site in 2003. They show the Packerton Building as I originally found it, in a state of abandon and dereliction.

These photographs are used to enhance the collection of illustrations used in this study.

They show how specific views of the actual building appeared in 2003, as compared to the illustrations, which shows the building when it was first constructed in 1905.

This building was completely demolished in 2007 leaving no trace of its existence.

Photo-1. This photograph shows the Packerton building as first viewed in 2003. What drew my attention to this structure was its size. From experience I knew this was an old building, but its lack of large wooden timbers normally used in construction during this time period indicated this was something new in the way buildings were to be built.

Photo-2. This is the south end of the building showing the south wall to the right of center and the long west wall stretching off into the distance. Measurements were taken of the entire structure to construct the computer models. Virtually every window was badly damaged or totally missing. Note the smokestack and the missing large doors.

Photo-2A. This is the south wall of the building. Note the window arrangement and the later modifications made at the right side of the south wall. The building was elevated sixty-four inches above the common ground level.

Photo-2B. This is the hinged door which was used to pass either mail or Bills of Order into the building. The fire alarm buttons were installed at a later date.

Photo-2C. Second floor windows in the south wall. The window on the left shows an interior wall of Stairwell-1.

Photo-2D. South wall in foreground, east wall is right of center. The Lehigh River flows southward past the building on the right of this photograph. A set of railroad tracks ran along the east side of the building at ground level to move material into or out of this warehouse.

Photo-3. This is the rear or north wall of the building. It had been modified several times. The illustrations show this wall as originally constructed.

Photo-3A. This doorway was the entrance to Stairwell-3. To the inside left is a small office with a window, followed by the staircase which is also on the left. At the rear of the right wall is the doorway that opened to the first floor.

Photo-3B. This image shows the doorway to Stairwell-3 and the small office window to the left of the doorway.

Photo-3C. In this photograph, the opening for the lift can be seen in the foreground ceiling. The window that was once in the center on the second floor had been removed, and the wall modified for a large doorway. On the first floor two sections of the north wall were knocked out.

Photo-3D. This view is from the rear or north wall towards the front or south wall. There is a doorway on the left which opens to the receiving area which is directly in the foreground. The rear wall of the receiving area has been knocked out. Above the receiving area is part of the lift opening in the ceiling to the second floor. The missing brick band on the right wall may have been where part of the lift mechanism was located.

Photo-3E. This view from the interior shows the north wall. The receiving area is to the left. The doorway in the foreground opened into the receiving area. The wall that was behind the receiving area was knocked out as were the two large openings in the north wall.

Photo-4. Smokestack next to the west wall near the south end of the building.

Photo-4A. This view of a part of the west wall shows the bottom section of the smokestack where it enters the base of the structure. The large door opening on the left is the first set of large doors encountered from the south end of the building which is on the right. Note also, part of the window on the right was bricked in at a later time.

Photo-4B. This photo shows one of the three large doors located in the west wall. Through this doorway, the doorway in the east wall of the building can be seen. The west wall and the east walls are similar in large doorway openings.

Photo-5. This photo shows how the base of the columns was mounted to a raised concrete base beneath the wooded flooring. It appears a layer of gravel and cinder was used to fill the areas between the column bases before the flooring was put down.

Photo-6. This image of former President Ronald Reagan was painted on the north wall of Stairwell-1 by an unknown artist sometime in the 1990's. In the foreground are the pipes and part of the cement structures that held the boiler for the air compressor that was added at a later time.

Photo-6A. This view is looking south towards the front or south wall of the building. On the right in the distance can be seen the portrait of the former President.

Photo-6B. This is the east corner opposite Stairwell-1. I believe this is the area where the air compressor was located. It shows the poured cement bases that once held the compressor. The wooden floors were removed in this area and a cement floor was added.

Photo-6C. This image taken from the corner of Stairwell-1 (on left) shows the base and other remaining concrete parts that supported the boiler. The image of Ronald Reagan is on the wall to the left of the corner. This view is towards the rear of the building.

Photo-7. This is the opening for the large double doors in the south wall. Note the concrete floor and the concrete fill between the metal plates. This concrete may have been added when the compressor was installed.

Photo-8. This view of the east wall shows the window and large door arrangement. In the foreground is the repaired wooden flooring and the central vertical columns.

Photo-9. This is looking into Stairwell-2 on the first floor. To the left is the doorway from the exterior. Directly ahead is a short hallway leading to a storage area on the right. Also on the upper right can be seen part of the molding which is on the bottom of the opening into the stairwell.

Photo-10. This is a view of the staircase leading to the second floor. The wall with the opening for the light into the staircase is on the left.

Photo-11. This view looking west shows the rear of Stairwell-2. It appears a wall was constructed on the right corner of the Stairwell, then later removed.

Photo-12. These are the steps within Stairwell-3 leading to the second floor. On the bottom left of this image was the doorway to the small receiving office.

Photo-13. This view shows how the ceilings were constructed and supported by the central columns.

Photo-14. Detail of the ceiling brick work. The hooks and eyelets shown were not used throughout the building but appeared at this location.

Photo-15. View of mixed floor boards.

Photo-16. View towards the north wall. Stairwell-2 is left of center.

Photo-17. View towards the north wall showing a thin track on the right with a wide separation between the remaining track further back.

Photo-18. A view looking down into Stairwell-1 from the second floor. The south wall with the window is at the rear.

Photo-19. The south wall through a fire wall doorway on the second floor. At the top of the image are windows that are in the clerestory south wall. The center tracks in the floor are at bottom center.

Photo-20. Windows in the clerestory south wall. Rivet heads can be seen in the column to the right.

Photo-21. View towards the north wall on the second floor. There is a fire wall in the foreground and one further back near the north wall.

Photo-22. This window is in the south wall of Stairwell-1. The walls of the building were all plastered on the interior with white plaster to increase the ambient light. In many places the plaster had fallen off the walls and ceilings.

THE CAR SHOPS OF THE LEHIGH VALLEY RAILROAD AT PACKERTON, PA.

The main car shops at Packerton Yards from "The American Engineer and Railroad Journal" dated 1894. This is just a small portion of the total railyard. The subject of this study was not constructed at the time this photo was taken.

ABOUT THE AUTHOR

I became interested in grist-mills and other types of water powered structures beginning in 1970. Grist-mills, as a structure, were in a sense complete flour making machines, and I wanted to learn what each component within them did and how they functioned.

I found the various types of water wheels and the gearing used to provide power to these mills very interesting. The development of the water turbine, though an ancient invention, eventually replaced the water wheel in mills and factories throughout the country.

The fact that flowing water was used as a power source was both a practical and common sense approach since the only other power sources then available were animal or wind.

The eventual transition from millstone grinding to make flour to the use of "Roller" milling was a further change in the evolution of flour making made during the late 1800's, and one that is still used to this day.

The development of steam powered locomotives, and other types of steam powered machinery during the mid-1800's continued the constant evolution of technological changes. Almost anything that provided some useful function or was a new invention during that time period I found interesting and worth investigating to add to my knowledge of that time.

The 19th century was a time of great innovation and invention.

It is that same curiosity that led me to the subject of this book. Why did this building remain the only building left standing at Packerton Yards before it was finally demolished in 2007?

I wanted to present this study because this building was unique at the time it was constructed and it changed the way industrial buildings in the future were to be built.

Lou Robertella-2016.